JOURNAL FOR THE STUDY OF THE OLD TESTAMENT SUPPLEMENT SERIES
250

Sheffield Academic Press

Beyond the River

New Perspectives on Transeuphratene

Josette Elayi and Jean Sapin

translated by
J. Edward Crowley

Journal for the Study of the Old Testament
Supplement Series 250

© 1991 Brepols for the original French edition (J. Elayi, J. Sapin,
Nouveaux regards sur la Transeuphratène, Turnhout 1991, coll.
Mémoirs premières).

Copyright © 1998 Sheffield Academic Press

Published by
Sheffield Academic Press Ltd
Mansion House
19 Kingfield Road
Sheffield S11 9AS
England

Typeset by Sheffield Academic Press
and
Printed on acid-free paper in Great Britain
by Bookcraft Ltd
Midsomer Norton, Bath

British Library Cataloguing in Publication Data

A catalogue record for this book is available
from the British Library

ISBN 1-85075-678-3

CONTENTS

ABBREVIATIONS

AASOR	Annual of the American Schools of Oriental Research
ADAJ	*Annual of the Department of Antiquities of Jordan*
AION	*Annali dell'istituto orientale di Napoli*
AJBA	*Australian Journal of Biblical Archaeology*
AmA	*American Anthropologist*
ANET	J.B. Pritchard (ed.), *Ancient Near Eastern Texts*
ANSMN	*American Numismatic Society Museum Notes*
AnSt	*Anatolian Studies*
AO	*Archiv orientální*
AOAT	Alter Orient und Altes Testament
ARAB	Ancient Records from Ancient Babylonia
BA	*Biblical Archaeologist*
Baghm	*Baghdaoler Mitteilungen*
BAH	Bibliothèque Archéologique et Historique
BARev	*Biblical Archaeology Review*
BASOR	*Bulletin of the American Schools of Oriental Research*
BCH	Bulletin de Correspondance Hellénique
BIOSCS	*Bulletin of the International Organization for Septuagint and Cognate Studies*
BIIRHT	*Bulletin d'information de l'institute de recherche et d'histoire des textes*
BMB	*Bulletin du Musée de Beyrouth*
CATAB	Centre d'analyse et du traitement automatique de la Bible
CIS	*Corpus inscriptionum semiticarum*
CUFr	Collection des universités de France
DAWW	Denkschriften der (k.) Akademie der Wissenschaften in Wien
IBS	*Irish Biblical Studies*
IEJ	*Israel Exploration Journal*
IJNA	*International Journal of Nautical Archaeology and Underwater Exploration*
IrAnt	*Iranica Antiqua*
JHS	*Journal of Hellenic Studies*
JJS	*Journal of Jewish Studies*
JNES	*Journal of Near Eastern Studies*
JNSL	*Journal of Northwest Semitic Languages*
JSOT	*Journal for the Study of the Old Testament*
KAI	H. Donner and W. Röllig, *Kanaanäische und aramäische Inschriften*

LCL	Loeb Classical Library
NAC	*Numismatica e antichità classiche*
NumC	*Numismatique chronicle*
OBO	Orbis biblicus et orientalis
OrAnt	*Oriens antiquus*
OTG	Old Testament Guides
OTSt	*Old Testament Studies*
PEQ	*Palestine Exploration Quarterly*
RA	*Revue d'assyriologie et d'archéologie orientale*
RB	*Revue biblique*
RBNS	*Revue belge de numismatique et de sigillographie*
RHPhR	*Revue d'histoire et de philosophie religieuses*
RN	*Revue numismatique*
RSF	*Rivista di studi fenici*
SEt	*Studi etruschi*
StIr	*Studia iranica*
TOB	*Traduction œcuménique de la Bible*
Trans	*Transeuphratène*
TynBul	*Tyndale Bulletin*
VAB	Vorderasiatische Bibliothek
VT	*Vetus Testamentum*
WVDOG	Wissenschaftliche Veröffentlichungen der deutschen Orientgesellschaft
ZAW	*Zeitschrift für die alttestamentliche Wissenschaft*
ZDMG	*Zeitschrift der deutschen morgenländischen Gesellschaft*
ZDPV	*Zeitschrift des deutschen Palästina-Vereins*

Preface

VISIONS OLD AND NEW

Ancient Vision

At the heart of the collective memory of modern Western society, a historiographical system, developed in the nineteenth century, focuses on and controls vast segments of the past. This plan has as its name 'The Origins of Christianity' and, like Janus, it has two faces. Before Jesus Christ, there would have been precursors of Christianity: Hellenistic Judaism and saving mysteries or religions, developed in various places in the Eastern world conquered long ago by Alexander the Great. After Jesus Christ, the conflict which up until then had set Hellenism and Judaism against each other would have been transformed. It would have become the conflict of the ancient church with the paganism and all the 'false' seductions of Graeco-Roman culture and its philosophico-religious syncretism. Scarred by the conflict, the church nevertheless emerged victorious, thus precipitating the decline of the ancient civilizations.

This framework still shapes the main elements of the vision that America and Europe share in the depths of their memory and place at the beginning of their common Western civilization.

New Fields, New Approaches

For several decades now, archaeology has considerably expanded the documentation on ancient Mediterranean and Near Eastern societies. The 'historians' history' has, at the same time, slowly won back its autonomy in relation to a self-assured philology which had for so long held it under its thumb. Historiography has begun to dismantle the network of obvious facts, embedded in obsolete topics whose recurrent questions have today lost all relevance. A redistribution of space has wiped out the old frontiers and opened up new fields. Finally, anthropology has seen its questions reach step by step all the sectors of human and social science. The specialists on ancient societies are already numerous enough to make anthropology the stimulus that their disciplines can utilize to explore the minor or major crossroads where societies and mentalities of the ancient world met, occasionally confronted each other, but always mutually influenced each other.

Mémoires Premières

Doing history thus takes on a new meaning, open to current events of the contemporary world. On the Persian Empire and the Hellenistic kingdoms which succeeded it, on the various movements in Judaism before and after the destruction of the Second Temple, on early Christianity and the emergence of new civilizations in the course of late Antiquity, on all these 'nodes' of remembrance and of still other related areas, novel views come up, innovative questions take shape. They herald insights that will replace those which are collapsing. A new report on antiquity is being born.

Translator's Note

As in earlier translations I owe much to the help and advice of Dr Pamela Milne, Professor of Hebrew Bible at the University of Windsor and must thank as well Professor Walter Skakoon of the French Department at the University of Windsor for his usual ready help.

Referrals to the Glossary are indicated by an asterisk.

A general, but not exhaustive, bibliography is included at the end of the book.
The citations of ancient texts and inscriptions follow the rules for transcription commonly used in each language. The biblical references are cited according to the *Traduction œcuménique de la Bible* (TOB).

We want to thank J. Briend, Professer at the Institut Catholique de Paris, A.G. Elayi, Maitre de Conférences at the University of Paris XIII, and T. Romer, Assistant at the University of Geneva, for the comments they provided us on the passages related to their respective specialities. We thank also Professor P. Davies at the University of

Sheffield, for publishing this English version in the *Journal for the Study of the Old Testament* Supplement Series, and Professor E. Crowley for his translation.

CILICIA

Tarsus

Haran

AMANUS

Mabbog

Al-Mina

Aleppo

Euphrates

Jebel Ansariyeh

Orontes

CYPRUS

Salamis

Kition

Aradus

Homs

Palmyra

Mount Lebanon

Byblos

Anti-Lebanon

Sidon

Tyre

Damascus

Jordan

Dor

Galaad

Samaria

Jaffa

Ammon

Jerusalem

Gaza

Lachish

Moab

Pelusium

NEGEV

Edom

Petra

ARABIA

SINAI

T. el-Kheleifeh

Map 1. TRANSEUPHRATENE

Climatic constraint: the pluviometric
limits of 200 mm (with variations)

a cycle of dry winters

a cycle of wet winters

0   100   200km

INTRODUCTION

In the history of the ancient Near East, a very promising new field of research has very recently opened up: Transeuphratene in the Persian period. Archaeologists invented the term 'Persian Period' as a handy way to designate the archaeological levels corresponding to the period of the Achaemenid Persian Empire.*

In the middle of the sixth century BCE, Cyrus II the Great (*Kūrush* in ancient Persian) revolted against his Mede suzerain, Astyages, and founded the Persian Achaemenid Empire. He then progressively enlarged it to the dimensions of the Assyro-Babylonian Empire and beyond, through a succession of conquests: first in 545 he annexed Lydia, the kingdom of the famous Croesus, and the Greek cities of the Asia Minor coast, then Eastern Iran, Syria-Palestine* and Cyprus, North Arabia; finally, in 539, he captured Babylon. The exact date of the conquest of Syria-Palestine and Cyprus is not precisely known and still remains very much under discussion.[1] The first campaign against Egypt took place in the reign of Cambyses (*Kambūjiya*), son and successor of Cyrus II; as a matter of fact, the Persians never succeeded in realizing a lasting conquest of Egypt. Having come to power by a coup, Darius I (*Dārayavahus*) passes for having been the great organizer of the Empire; his reign was especially marked in the West by the conflict with the Greeks and the first Median war in 490. His son, Xerxes I, continued the administrative work of his father and confronted the Greek cities in the second Median war. During the reign of Artaxerxes

---

1.   According to Herodotus (*The History*, I.143), at the time of the conquest of Lydia, 'the Phoenicians were not yet subject to the Persians'; Cyrus declared for his part that all the kings of the western states had come to make their submission at Babylon (*ANET* 316a). The generally accepted date varies between 545 and 539, but the low dating—526/525—which had had few defenders has just been defended anew (Watkin 1987), without so far producing conviction; there is no reason at present to abandon the traditional early dating, unless it becomes possible to be more precise (Elayi 1989b: 137-38).

I the conflict with the Greeks continued, to which was added among others the revolt of Megabyzos in Syria. The reign of Darius II seems to have been marked, in the western part of the Empire, by a relative peace that allowed the power of the Satraps to be consolidated. The fourth century was a period of successive disturbances in the region, characterized by several rebel movements of local and satrapal authorities, under the reigns of Artaxerxes II, Artaxerxes III, Arses and Darius III. With a victory over the troops of Darius at Granica in 334, Alexander the Great brought an end to two centuries of Persian Achaemenid domination by successively occupying the different possessions of the Empire.

Despite the recent progress in research on the history of the Persian Achaemenid Empire, its domination over Transeuphratene is still quite poorly known and a research effort in this area should be implemented, because it is indispensable to know the narrow setting into which the local politics of the States and provinces of the region had been fitted. It is important, too, to characterize the period of Persian domination in comparison with the succession of foreign dominations in the Near East—that of neo-Assyrian and neo-Babylonian Empires, then that of Alexander and his successors—by bringing out the continuities and the disruptions.

What exactly do we mean by the term 'Transeuphratene'? The answer is not simple, since it has been used to designate several geographical entities according to the period, the sources and the usage of the author who referred to it. Generally, however, we can say that this term designates the region situated 'beyond the Euphrates', namely to the west of this river from the perspective of an observer situated to the east, in Mesopotamia or on the Iranian plateau for example. It already existed in Akkadian* in the time of the neo-Assyrian Empire, where it is found for the first time, under the form *Eber nāri*, in the letter texts of the reign of Sargon II (Parpola 1987: n. 204, r. 10): it designated in that case the Syro-Phoenician regions west of the Euphrates. In the treaty concluded between Assarhaddon and King Baal of Tyre, there was an invocation of the gods of the three principal regions of the Assyrian Empire: those of Assyria, those of Akkad, and those of Transeuphratene (Pettinato 1975: 152-54, col. IV.1.1-20). In the *Annals* of Assurbanipal, there is mention of Transeuphratene in a list of royal vassals: 'The great cupbearer (*rab shaqē*), the governors, the kings of Transeuphratene, all my vassals (ARAB, II: 901).' However, when the term comes up in biblical Hebrew under the form *'eber han-nāhār*, it designated, in contrast with the Akkadian perspective, the region situated east of the Euphrates (2 Sam. 10.16; 1 Kgs 14.15). But the Akkadian perspective persisted in the Aramaic of the Empire, since *Abar-naharā* is the form used in the Aramaic section of the biblical book of Ezra (Ezra 4.10, 11, 16, 20; 6.6, 8, 13). In Greek, this term is met under the form *Peran Euphratou* or *Peran tou Potamou*, with the same meaning: thus, in the reign of Demetrios I Soter (about 162–149), the control exercised by Bacchides, one of the 'friends of the king', over Transeuphratene involved the region situated between the Euphrates and the Egyptian frontier (1 Macc. 7.8; cf. 3.32). No matter what its form, this term always expressed the point of view of the central authority of the empires that successively governed this region.

In the Persian period, Transeuphratene made up the largest territorial and administrative division of the Achaemenid Empire. It represents in our eyes a quite poorly defined entity, which, oddly, seems to have been called 'Assyria' in Old Persian (*Athurā* or *Athuriya*) and in neo-Elamite (*Ashshuraip*).[2] A number of problems come up in regard to its nature, its unity, its evolution, its organization and its limits.

First of all, it is very hard to say by what term the Persians characterized the status of this area. Most modern authors do not see the difficulty and use the term 'satrapy', of which they make an ill-considered use to designate all kinds of Persian districts. Now, there is no need to confuse 'satrapy' and 'province': a province was a territorial subdivision of a satrapy, often corresponding to a geographical, ethnic and linguistic entity, with political structures often inherited from preceding Empires; we do not know the number of provinces forming part of Transeuphratene. The term 'satrapy' comes to us from Greek, but even the contemporary Greek historian, Herodotus, seems not to have had an altogether clear idea of the nature of such an administrative area. This is the way he writes in regard to Darius I: 'He established in the Persian Empire twenty governments, which they themselves called satrapies' (Herodotus, 3.89). He introduces the term 'satrapy' as if it were Persian and uses, for his purposes, either *arkhē** or *nomos**, with Transeuphratene constituting the fifth *nomos*, which paid a tribute of three hundred and fifty talents* of silver; he apparently assumed the

---

2. Inscriptions *DSf* l. 25-28; *DSz* l. 26-29, Vallat 1972: 3–14. On Achaemenid Transeuphratene, see also Rainey 1969; Calmeyer 1990; Lipiński 1990.

coincidence of fiscal areas and governments or satrapies. But it is surprising just the same that the term supposedly used by the Persians—
'satrapy'—is missing from all the Achaemenid inscriptions so far
known. The Imperial lists use the term *dahyu*, whose meaning is still
very much under discussion (Herrenschmidt 1976: 62-63; Lecocq
1990), but its ethnic aspect seems to have taken precedence over the
administrative aspect. On the other hand, the term 'satrap' existed,
under the Old Persian form *khsathrapavan*, in the Behistun inscription, and under some semitizing forms in the Aramaic of the Empire
and in Akkadian;[3] but it could have designated just as well the governor
of a large district like Transeuphratene as well as one of its subdivisions.

We still do not know to what extent Transeuphratene constituted at
that time a territorial, administrative and fiscal unity, since its boundaries are elusive and underwent changes in the course of the two centuries of Achaemenid history. In fact, at the beginning of the Persian
empire, Cyrus the Great combined Transeuphratene and Babylon under
the control of the same satrap, who was Iranian. The first one, Gobryas
(*Gūbaru*), filled this office from 535 to 525 (the fifth year of the reign
of Cambyses) and perhaps even to 521, with the title of 'governor of
Babylon and Transeuphratene' (Leuze 1935: 36-37; cf. Joannes 1990;
Stolper 1989: 290-91). The second, Hystanes (*Ushtānu*) (Herodotus,
7.77), had the same title in documents dated from 521 to 519 (year 1
and year 3 of Darius I; Stolper 1989: 290-91 and bibliography); he
seems to have remained in office from 521 to 516, and perhaps even
later. There is no agreement on the date and circumstances of the separation of Transeuphratene and Babylon. Three principal hypotheses
have been proposed: according to the first, the separation would have
taken place at the latest in 503 since a document dated the nineteenth
year of Darius (503) mentions a certain *Bagapa'*, 'satrap of Babylon'
(Stolper 1987: 396; 1989: 292 and n. 9), and another text dated to the
twentieth year of this same king mentions *Tāttanū* (Ungnad 1940–41:
240-43) (no doubt the Tattenai of Ezra 5.3, 6; 6.6, 13), who was 'Satrap
of Transeuphrates'. This first hypothesis seems to be contradicted by
the publication of a new document dated to the thirty-sixth year of
Darius (two months before his death), which mentions Huta (*Ḥuta?*),
'governor of Babylon and Transeuphratene'.[4] Since the term 'satrap'

    3.   Weissbach 1911: 44-45, §38 (Old Persian); Lipiński 1975: 197-208, l. 5
(Aramaic: *ḥshtrpn'*); McEwan 1982, no. 48, 6 (Akkadian: *aḥshadrapānu*).
    4.   Stolper 1989. We draw attention as well to the hypothesis presented recently

had an extensive semantic field, *Bagapa'* and *Tāttanū* were perhaps subordinate satraps, in charge respectively of Babylon and Transeuphratene and answerable to a satrap-in-chief who still supervised the two regions. The second hypothesis connects the separation of Babylon and Transeuphratene to the Babylonian uprisings which took place at the beginning of the reign of Xerxes (486) (Stolper 1989: 292 and nn. 10-11 [with bibliography]): this connection is possible, but seems unnecessary. The third hypothesis, the most cautious, proposes a more or less lengthy period, which extended at most from the beginning of the reign of Xerxes I (486) down to 420 BCE, a date at which the governor of Babylon had the simple title of 'governor of the land of Akkad' (Stolper 1989: 293 and n. 12 [with bibliography]; 297-98). In the fourth century, Mazaios (*Mazday*) brought together under his control Cilicia and Transeuphratene; however, he never had, as far as we know, the title of satrap, but that of 'attendant to Transeuphratene and Cilicia' (Babelon 1910: 457-62).

The boundaries of Achaemenid Transeuphratene are very difficult to establish, all the more so since the idea of a political 'frontier' was very different from our modern notion of frontiers between nations: the reality suggested by this term was at that time neither fixed nor precise and uniform; it varied with the times and political events, and could include autonomous countries with their own frontiers, like the Phoenician cities and their possessions, which did not depend directly on the central government. The eastern frontier seems to have been the Euphrates, since Babylon formed a specific province; the northern frontier passed somewhere through the mountainous region of the Amanus, which separated Transeuphratene from the neighbouring area of Cilicia, and went as far as the mountainous massif of Ṭūr 'Abdīn, situated at the source of the Habur; to the west, it did not stop at the sea, since it included Cyprus according to Herodotus (Manfredi 1986; Lipiński 1990; Herodotus, 3.91). The southern frontier is quite uncertain for two main reasons: on the one hand we do not know the exact extent of the territory of the Arabs and whether it was included in Transeuphratene or formed a distinct area (Lemaire 1990; Knauf 1990); it was possible that the Arab kingdom of Qedar included, towards the north, Gaza, Lachish, Maresha, Hebron and Ein-Gedi, at least up to the time of Nehemiah and even up to 400 (Lemaire 1990);

by Lipiński 1990: after having been separated for some years, Transeuphratene and Babylon would have been once again reunited and then later definitively separated.

another possibility is that the Qedarite kings had obtained ordinary commercial privileges in the southern part of Palestine, without any political implications. On the other hand, during the first two-thirds of the fourth century, that region seems to have undergone profound changes in its political organization, as we will see in Chapter 9.

What was the capital of Achaemenid Transeuphratene (Elayi 1989b: 144-46)? Several hypotheses have been proposed and none of them is really convincing. The Sidon hypothesis is based on several arguments: the political importance of that city among the Phoenician cities; the primordial role of its fleet in Persian maritime strength; its geographical position as an ideal operational base; iconography of Persian inspiration in its coinage; introduction of Persian style of dress at its court, and, finally, the presence in the city of a Persian residence and *paradeisos**. But the text of Diodorus, which mentions the presence of a satrap at Sidon, is not a decisive factor since the author obviously did not know what a satrap's functions were. The Tripolis hypothesis is based on a faulty interpretation of a passage from Diodorus and on the term *tarp^elāyē'* which is found in the difficult verse, Ezra 4.9, in which some have seen the chancery official of Tripolis (Diodorus, 16.41.2; cf. Elayi 1990b): even if that were the case, which has little probability, it would imply the presence of a Persian provincial administrative centre in the city, but not necessarily the presence of the satrap of Transeuphratene. The hypothesis of the springs of Dardas (Sínek?), between Aleppo and the Euphrates, is based on a passage of Xenophon that situates the Persian palace of Belesys (*Bēlshunu*) there (Xenophon, *Anabasis* 1.4.10; Manfredi 1986). As for the Damascus hypothesis, it relies on allusive classical sources (Strabo, *Geography* 16.2.20; Quintus Curtius, *History* 4.1.4); on its central position in Transeuphratene; on the fact that Darius III seemed to have deposited his treasure there and on the fact that it could have been capital of a satrapy after Alexander's conquest. As a matter of fact, the problem has been poorly presented, since Transeuphratene did not necessarily have a unique or fixed capital, if we bear in mind its expanse and the reorganizations it went through. Thus, at the time of its revolt, about 350 BCE, Sidon perhaps had a temporary satrapic residence because of Persian preparations against Egypt, but no satrapic army since the Sidonians were able to prepare their revolt without having to worry about a Persian garrison in the city; at the approach of Alexander, the existence of a satrapic residence at Damascus was not inconceivable, since that city was certainly more secure than Sidon because of its geographical situation, which made access more difficult.

If Transeuphratene still remains so difficult to define, it is because this is a region that for several different reasons has been misunderstood. The quantity and importance of the data that we have at our disposal are very variable as we go from sector to sector: we can contrast, for example, the relative richness of the information on Judaea and its scarcity in regard to the Negev, while noting however that this drawback is not specific to Transeuphratene. On the whole, the data are not as plentiful as for other regions of the Persian Empire and as for this region in other periods, but all specialists in antiquity know that the abundance of the data they deal with is always very relative. In reality, they are rather poorly known in this case due to their dispersion, their diversity and the lack of interest in that epoch.

The archaeological exploration of Transeuphratene comes up against a whole series of difficulties, of which some affect the Persian levels only: thus, these levels have been quite frequently damaged or even destroyed by the Hellenistic and Roman levels, which were better structured. As a general rule, the archaeological exploration remains unbalanced, since the sites in the hinterland have been much less excavated than those of the coastal strip; a good number of them, however, are covered by modern built-up areas and cannot be sites for excavations. Neither can we underestimate the consequences, for archaeology, of the present political situation in the Middle East: namely, the difficulty of exchanges between researchers in the region; the a priori ideological differences in considering the past; the impossibility of excavating certain sites; the increase in the ravages perpetrated by clandestine excavations.

Many difficulties must be overcome, then, by researchers who are working in this new field of research represented by Transeuphratene in the Persian period, but the exceptional interest in this section of regional history constitutes for them the best of stimulants. In fact, quite remarkable evolutions and cultural changes took place at that time in this region and they call for study: a key stage in the redaction of biblical texts belongs in this period; in addition, we see the introduction of money and the origin of the cash economy, with a change in economic attitudes; the sources of tensions connected to problems of power and identity; the emergence of city-states similar to the Greek-city type;

the development of mercenary armies; the opening up of the Western
fringe of the Persian Empire to the Greek world, and so on.

In this work, we will propose for the readers a new look at Achaemenid Transeuphratene: our objective is actually to make known to
them this new field for research in all its diversity, with its numerous
implications and its potentially immense riches; to define the conditions
for a fruitful exploration and the short- and long-range objectives;
and to present some of the promising first results already obtained
with this new perspective. Without polemical intentions of any kind, it
seems, however, essential to stress the methodological errors already
committed or liable to be committed in this field, so that research can
develop on more scientific foundations.

Chapter 1

Traditional Approaches

If research on Transeuphratene in the Persian Period has until now
been neglected, this is not only because of the real difficulties that it
presents, but also because of various errors about its value, of which
the two principal ones are the skewing of archaeologists' interest and
the development of derivative studies in certain areas.

For a long time, in fact, the Persian levels have remained outside the
real interests of archaeologists. For those in the ancient Near Eastern
field, these levels have the drawback of being too recent, since one of
the first criteria in their eyes was the criterion of antiquity. Such an attitude was an offshoot of spectacle-archaeology, whose principal objective was to magnify the first great civilizations or, quite simply, to
search for the sensational. The syndrome of going back in time, which
has been so popular in the comic strips, expresses in fact the preoccupations of the modern world which have excited several generations of
researchers—namely, the search for the origins of civilizations. This
has always met quick success with the general public, relentlessly in
pursuit of the enigmas for which, all things considered, it provides
the key. We still remember the excitement over such books as S.N.
Kramer's *History Begins at Sumer* (it was published in French translation in an Arthaud edition in 1957): the 'Greek miracle' had had a
precedent, the Sumerian civilization, born some 5000 years ago in the
south of present-day Iraq! The acknowledged objective of this kind of
work was to recover the 'initial hearths. . . from which immense regions have been set ablaze', and those who, 'in opening up a path for
humanity, showed it the summits' (Parrot 1981:51).

The site of Ras Shamra (Ugarit), on the Syrian coast, has been the
scene of some fifty excavation campaigns by French missions, directed
in turn by C.F.A. Schaeffer (1929–69), H. de Contenson and J. Mar-
gueron (1970–76), and M. Yon (beginning in 1978). These campaigns

have progressively revealed a brilliant civilization of the second millennium, the Ugaritic civilization, which vies with Byblos for priority in inventing the alphabet. Ugaritic studies have seen such a development that they have become a separate research area, in which several dozens of researchers now specialize. No matter how remarkable the results obtained at this one site, they cannot justify the fact that the installations from the Persian period have been neglected, even if it was just a simple country settlement. This was plain negligence, obviously, since this installation at the summit of the tell and on the surface has been cut across by the archaeologists in a hurry to reach the levels richer in remains and rewards. We had to wait until the thirty-second excavation campaign in 1971 for a more careful exploration of the Persian level to be undertaken: we should consider ourselves happy to have R. Stucky's book, which presents its results and integrates into it a reevaluation of some of the elements previously published (Stucky 1983). We could multiply examples of this kind: at Jbeil (Byblos) on the coast of Lebanon, P. Montet was only interested in the levels corresponding to the period of Egyptian domination (Montet 1926); we had to wait for the excavations of M. Dunand, especially beginning in the 1960s, for the Persian levels to be gradually revealed and, as such, taken into consideration (Dunand 1964: 33-34; 1966: 97-101). The considerable excitement about the discovery of third-millennium texts at Tell Mardikh (Ebla) in the north of Syria, and the specialized international debates on the identification of the language of Ebla, had let the excavators ignore for a long time the fact that there was also an installation from the Persian period on the site: the recent publications of S. Mazzoni have just revealed this (Mazzoni 1990).

For the archaeologists of the Classical world on the other hand, the Persian levels were too old. A site was worth the trouble of being excavated to the extent that it possessed an installation later than the period of Alexander's conquest. In this perspective, it mattered little that there were not older levels; ultimately, the historical interest in the site was even intensified when there was a Hellenistic or Roman 'foundation'. We may cite the case of Baalbek (Heliopolis), in the Lebanese plain of the Beqaa Valley, a Greek city which became a Roman colony under Augustus, and underwent a development under the Antonines: its famous cult of Jupiter Heliopolitan makes one forget a bit too much the earlier cult of a Semitic Baal of which the name of the site has preserved some traces. It is scarcely remembered either that the rich

Caesarea of Herod the Great, whose importance increased still more with the destruction of Jerusalem in 70 CE, was originally a Phoenician site, known in the classical sources as Straton's Tower (Levine 1973; Ringel 1975).

Archaeologists of the Classical world are sometimes interested in levels earlier than the arrival of Alexander, when they suspect that a site might have been a pre-Hellenic Greek settlement. Several excavations have been undertaken in the Near East in order to study Greek settlement in the region before the Hellenistic period; the older levels have then been taken into consideration to try to date the arrival of the Greeks on the Levantine coasts. Beyond the Greek 'colonization' of the first millennium, they are delighted to be able to go back as far as the Mycenaean and even Cretan 'colonization'. When L. Woolley embarked on the excavations of Al-Mina on the Orontes and of Alalakh (Tell Atchana to the South of the Lake of Antioch) from 1936 to 1949, his avowed objective was to find pre-Hellenistic Greek settlements (Woolley 1964: 10-11; 1938: 1): 'the object in view was to trace connexions between the Middle East—Mesopotamia, Anatolia and Lake Van included—and the Aegean' (Woolley 1964: 10-11). According to Woolley, the history of Alalakh 'bears on the development of that Cretan art which astonishes us in the palace of Minos at Knossos, it is associated with the Bronze Age* culture of Cyprus, bears witness to the eastward expansion of the trade of the Greek islands in the protohistoric age, throws an entirely new light on the economic aspects of the Athenian empire' (1938: 1). The results of his excavations were distorted by that a priori approach (see Elayi 1987a: 249-54: thus, at Al-Mina, he discovered ten levels, of which the oldest, lying on virgin soil, dated from the eighth century; but starting from the idea that the site was older, he ventured hypotheses to explain the absence of earlier levels; on the other hand, the oldest levels and the Greek material were favoured to the detriment of recent levels and non-Greek material.

If the sidetracking of the interest of the archaeologists in relation to our field of studies is an old phenomenon which seems on the way to dying out,[1] the drifting off of studies in certain areas is on the other hand a recent phenomenon that tends instead to be on the increase. The example of Phoenician studies is one of the most typical (Elayi:

1. We note however that the *IVᵉ congrès d'histoire et d'archéologie jordanienne*, which was held at Lyon from 30 May to 3 June 1989, provided no separate space in its chronological programme for the Persian period.

1990a): the pretext put forward is the political situation in the Near East the last fifteen years or so which reduced the possibilities of archaeological exploration in Levantine Phoenician sites. We have observed during that time, under the form of a drift toward the west, the shifting of the areas of interest of Phoenician archaeology; as if the Phoenicians had transmitted to modern researchers the taste for travel, they are making little by little the tour of the Mediterranean: first,Cyprus, then Italy, Tunisia, Spain, and perhaps next Morocco. We can measure this drift perfectly in the publications, the colloquia, the congresses and expositions:[2] we are reaching at present the 'Spanish phase', and the expansion continues. This drift has grown to such an extent that, sometimes, Phoenicia is not even mentioned any more: have they forgotten the cities of Tyre, Sidon, Byblos, Beirut and Tripolis in Lebanon, and Arwad in Syria? Phoenician history, reduced to its colonizing phase, is on its way to becoming an exclusively maritime history. The real renewal which maritime history has undergone during recent decades through its opening up to the new problematics of economic and social history, historical anthropology and the history of attitudes is probably not a stranger any longer to this transformation. This renewal owes much to *La Méditerranée et le monde méditerranéen à l'époque de Philippe II*, by F. Braudel (1966; 1969: 11-38), which was presented as a project of total history, focused on marine space: what more tempting than to write the history of a people who were always considered exclusively a nation of seafarers? And this is not taking into account the fact that the maritime history of the Phoenicians, thought of in this way, could easily be popularized for our generation because of the fascination exerted by the sea as a symbol of travel, adventure and discovery; this would explain, at least in part, the success for some years now of cultural events and popular publications on the Phoenicians;[3] after having aroused since the pioneering work of Renan alternating waves of 'Phoenicia mania' and 'Phoenicia phobia', the Phoenicians, as sea adventurers, find great favour with the public, nourished by books such as those of S. Moscati for example, *The World of the Phoenicians* (1968) or *L'enigma dei Fenici* (1982).

2.   If we prepare a balance sheet on the second International Congress on Phoenician and Punic Studies which was held at Rome in 1987, we find that of 150 papers, only about 30 were about the Phoenicians of Phoenicia.

3.   The exposition 'I Fenici' which was held from 6 March to 6 November 1988 in the Grazzi palace at Venice, had an unprecedented media success.

Far be it from us to underestimate the interest in maritime history and the phenomenal contribution of publications on the Phoenician–Punic sites of Cyprus and the west; in the same way, we must express our satisfaction with the media interest from which Phoenician studies profit today. But we cannot accept the truncating effect of their drifting away, as much on the historical vision that it gives the general public as on the development of research in this field. Let us put in its true place the maritime history of the Phoenicians, which is only one aspect of their history, no more and no less; on the other hand, despite the interruption in excavations in Lebanon, there still remain enough unexploited data, and there are sufficient Phoenician sites, in the process of exploration or worth exploring, in the north and south of Lebanon to allow for continued research.

We meet another example of drift in studies on the Achaemenid Persian Empire. Once again recently, Syria-Palestine was listed among the many 'unknown lands' of the Empire, certainly full of promise, but without the possibility of immediate exploitation, with the three supports of the new Achaemenid history being the coast of Asia Minor, Babylon and Egypt (Briant 1988: 138), because of their rich documentary records. Without any claim to the same richness, the records of Syria-Palestine are, however, starting to expand and especially to become exploitable: not taking into account all that is presently available there would constitute, in our view, a methodological error.

When it has not been completely neglected, the area of research on Transeuphratene in the Persian period has often been insufficiently or poorly investigated by what we will call, to simplify matters, the 'traditional approaches', namely, the practice followed by researchers of an era now past and one which is still followed by some modern researchers. The first was quite respectable, since each age can only use the means at its disposal, and every researcher today is conscious of what is owed to the remarkable work achieved in the past, relating closely or remotely to Transeuphratene in the Persian period: that of Renan, the founder of Phoenician archaeology, and of his disciple Ch. Clermont-Ganneau, for example.[4] The second practice, adopted by some modern researchers, aiming at exactly following in the footsteps of their predecessors, represents a commendable expression of fidelity, but is at the same time an easy way out and a brake on research: as such, it should be

---

4. Cf. Gran-Aymerich and Gran-Aymerich 1987a and 1987b, for a brief survey of the personality and work of these scholars.

exposed, with all the more vigour when the researchers involved are responsible for research projects and training. The two main weaknesses of this approach lie in the narrowness of the views and its a priori ideas.

The narrowness of views is sometimes found in unidisciplinary studies. Even if the author indicates in an unassuming way that he approaches a question from the sole point of view of his speciality, the integrity of the procedure does not for all that make up for its weakness: the unidisciplinary approach to a question can distort its study, without the author even being aware of it. As F. Braudel wrote, 'Each specialist has pieced together the puzzle in his own way... In brief, one fact is evident: each social science is imperialistic, even if it denies that it is so; it tends to present its conclusions as a global vision of humanity' (Braudel 1969: 86). Political history, a form of history inherited from Graeco-Roman Antiquity, can illustrate this type of approach, since it subsisted exclusively from unique and singular historical facts, the events, and has but lately been denounced for its pretension to primacy and its exclusivism. Nevertheless, notwithstanding the attacks of the Durkheimians since the end of the nineteenth century and the development of History of Long Duration and of Seriality*,[5] political history has not disappeared; on the one hand, it remains the history meant for the general public and, on the other hand, it has been rehabilitated by historians at the cost of a real mutation, by profiting from new orientations and methods tested in the study of economic and social phenomena. Contemporary historians, faced with the permanent amplification of events by the media, have been obliged to rehabilitate their study, as a reflection of the structures of a given society and an agent of their evolution (Nora 1974; Rémond 1982). To cite an example in our field, the linear account of the taking of Tyre by Alexander the Great, which appears in tragic and spectacular style in all the political histories of the Near East, could profitably be taken up again today as a sign, an observation point and a revealer of Tyrian society in 333 BCE, seen through the knowledge, mentalities and motivations of the different classical authors concerned; through these we can also grasp

5.   In particular with M. Bloch, L. Febvre and F. Braudel: Braudel 1969: 41-56, 135-154; Febvre 1953: 61-74, 114-18, who wrote nevertheless: 'There is no reason ever to discourage people from learning political history. And I do not approve either the sense of disdain that some profess in dealing with "factual history" and "historicizing history".'

in part the impact of this major event on the evolution of the structures of that society.

An inadequate approach, too, is the selective approach, which gives greater importance to the study of the ruling social levels: for example, in political history, the lives of the 'great men'; in religious history, the great divinities protecting the cities; or, in archaeology, the palaces and the temples. The motivations are of little importance, whether elitism, a taste for the spectacular, or an easy answer, since the data in this domain are most accessible and abundant. Without denying the usefulness of this type of approach, we will emphasize the need to complete it, in the examples mentioned, by looking at the evolution of society in its entirety, by the study of popular devotions and by a 'domestic' archaeology. It is very interesting to learn about the temple and cult of the 'Lady of Byblos' (*Ba'alat Gubal*), the dynastic divinity of *Yeḥawmilk* in the fifth century, but it would be at least as important to know more about the religious beliefs of simple residents of Byblos: what gods and goddesses did they invoke and in what circumstances? What rituals did they carry out and with what motivations? What powers did they attribute to such or such a type of amulet?

The narrow views of the traditional approaches also find expression through the partitioning, and then the hierarchy, of disciplines and areas of research. Without prejudging what happens in other sectors of research, we still today note, in ours, the survival of a certain sectarianism of disciplines, which can go from real or feigned ignorance to latent or open confrontation, between the 'corporations' of archaeologists, epigraphers, exegetes or historians, for example. As a consequence of this sectarianism, we sometimes even see the development of a real professionalism in systematic and negative criticism, which usually conceals the fact that the researchers involved have nothing to say outside their narrow speciality, nor even at times anything further in their own special field or one alleged to be such;[6] we cannot confuse, however, this condemnable form of criticism with a non-polemical and constructive criticism, which is always useful and stimulating for research because it generates an authentic dialogue. Already in 1900, the historian M. Clerc indicated the absolute necessity of dialogue between archaeology and other disciplines: 'archaeology cannot

6. Langlois and Seignobos, even at that time wrote: 'It is excess in criticism that succeeds, as well as the crudest ignorance, which leads to error . . . Hypercriticism is to criticism what trickery is to subtlety' (1898: 107).

and should not be sufficient unto itself... Left to its own resources, deprived of the help of other auxiliary sciences and especially, of the general data of history, archaeology would risk ending up, despite exact observations, with radically false conclusions.'[7]

Since then, interdisciplinary cooperation has become the leitmotiv of research on Antiquity. The promoters of the 'new history', among others, have based their innovating procedures in part on a rejection of the partitioning between the social sciences and on making evident the multiple interactions of the 'builders' of the historical movement.[8] And yet we still meet today as an experiment, limited to the study of Syria in the Bronze Age (the Persian period will come later!), the setting up of a dialogue between two neighbouring disciplines: epigraphy and archaeology (Gates 1988). The experiment with 'tandems', associating for example a Hellenist and an Orientalist (Baslez and Briquel 1989), certainly is not without interest, but remains selective and limited, because of the specialized character of the fields that are brought together, and cannot lead to a real interdisciplinarity. Our intention is not in any way to criticize such endeavours, since they are necessary, but we wish to emphasize to what extent they reveal the fact that interdisciplinary research in our field remains most of the time just a pious wish. All in all, the situation has changed but little since Braudel wrote in 1960: 'Nothing proves better that sort of present irreducibility of the social sciences one to the other than the attempted dialogues, here and there, over the frontiers' (repr. Braudel 1969: 88).

The hierarchy of disciplines is perhaps one of the main reasons for the persistence of their compartmentalization: it so happens that each specialist considers his discipline as superior and at times even complete in itself, so much so that research can find itself shut up in a Pirandelline situation of 'To each his own truth'. Some disciplines, however, are fairly unanimously considered to be ancillary: such is the case with numismatics, which L. Mildenberg recently emphasized by beginning in these terms his paper on numismatics at our first interdisciplinary Colloquium on Syria-Palestine in the Persian period: 'At the latest international congresses and symposia, a lot has been said about the importance of interdisciplinary research, but all we realized there

7. Clerc 1900; Garelli 1964: 123, spoke of 'the absurd dichotomy between archaeology and epigraphy' in Assyriology.
8. See, for example, Lefebvre 1978; 1980. A colloquium on interdisciplinarity was again held 12–13 February 1990 in the Palace of UNESCO at Paris.

was the fact that numismatic study was considered the indispensable, but despised handmaid, the classical *ancilla*, but not an essential discipline' (Mildenberg 1990: 137-38).

There exists as well a hierarchy of research fields, with some being considered more important than others, often according to very debatable criteria.[9] Thus, the study of Transeuphratene in the Persian period has up until now hardly been considered a priority area: for certain Orientalists, the Persian Empire is a 'peripheral' Empire and for certain Hellenists, Alexander has not yet arrived! The compartmentalizing of areas of research is perhaps still more pronounced than that of disciplines, and isolates researchers to a greater extent, setting up blocs that are most often narrow: the exegetes of the Old Testament, specialists of North-West Semitic*, those of the Classical world, those in Cypriot studies, the Iranists, Assyriologists. The 'bridges' are still rarer between disciplines: thus, the collaboration of A. Lemaire, a specialist in North-West Semitic, and J.M. Durand, an Assyriologist, in *Les inscriptions araméennes de Sfiré et l'Assyrie de Shamshi-ilu* (1984), was probably so exceptional that it was emphasized at length in the Preface and described as 'propitious and exemplary'.

We can ask what is the place of history in such a dividing up and compartmentalizing of disciplines and fields: it is at the same time everywhere and nowhere. It is everywhere in this sense that most specialists treat it as a second-rate discipline and consider themselves as being also (in an accessory way) historians, insofar as they study the questions that come up through their specialities, often very specialized ones; in the best of situations, they become initiated in one or two other disciplines or collaborate with one or two other specialists. One quite often ends up in this way with a segmented, sectored history, without any real historical problematic being proposed. In the following chapter we ask what should be, at least in theory, the role of the historian of Transeuphratene.

The traditional approaches are cluttered up with some aprioristic approaches, especially 'biblocentrism', 'hellenocentrism' (Elayi 1987b: v), and very recently 'Iranocentrism'. Biblocentrism is as old as biblical studies: it consists of studying all questions, even those which are completely independent of the Bible, through a biblical perspective. In other words, in this perspective, the world of the Diaspora of the

---

9.   (Braudel 1969: 95) wrote: 'I hold that for a unity to be constructed all research is significant'.

ancient Israelites, from the neo-Assyrian and Babylonian deportations
to Mesopotamia, from the end of the eighth century to the beginning
of the sixth, and their dispersal from Upper Egypt to Susa in the setting of these Empires and the Persian Empire, is considered the 'biblical
world'; all of Palestine, the Philistine and Phoenician coasts, in addition to the ancient land of Israel and Judah, become the 'land of the
Bible', the 'Holy Land'. How many publications on the region have
included and still include in their actual title an expression of this kind!
So we have the invaluable archaeological synthesis of M. Avi-Yonah,
Encyclopedia of the Holy Land, published in 1975–78, and more recently *The Archaeological Encyclopedia of the Holy Land*, edited by
A. Negev (1986). In the same way, the archaeology practised on numerous sites of Syria-Palestine has long been a 'biblical' archaeology, in
this sense that the stratigraphy* was frequently established from the
great stages defined in the Old Testament: for example, pre-exilic or
postexilic levels. Most of the place names mentioned in the Bible have
become the place names of the 'biblical world', so that archaeological
exploration has often had as an objective to localize them and many
sites have been excavated with this perspective in view, which, however, has been something positive, since without this motivation they
would probably still not be excavated.

Among the disciplines and fields annexed by biblical studies, we will
especially mention Phoenician studies which are striving today, with
difficulty, to gain their autonomy: if the maritime history of the Phoenicians has to a great extent managed well in this enterprise, studies on
the Phoenicians of Phoenicia are not yet really liberated.[10] Thus, when
biblical scholars, exegetes or theologians, study a specific problem of
Phoenician history, they do not do it for the interest it presents, but to
elucidate a problem posed by an Old Testament passage. The biblical
problem usually happens to be resolved, but more rarely the Phoenician problem under consideration, which is there only in an Old Testament perspective. It is altogether legitimate that biblical scholars study
the environment of the Old Testament, but it is inadmissible that they
annex the disciplines relative to what represents in their view that environment:[11] each discipline deserves its autonomy and has real scientific
value only at this price.

10. They are included, for example, in *ANET* and *ANEP*.
11. To cite a parallel example, Briant 1982: 302 n. 43 and 304 n. 50 notes that there

Hellenocentrism is a little more recent distortion than biblocentrism, but just as harmful for studies on Transeuphratene. J.G. Droysen, with his *History of Alexander the Great*, which was published in Germany in 1833, seems to be at the origin of the concept of the cultural superiority of Hellenism, from which Hellenocentrism is derived. Up until then, the restrictive judgments on the morality of Alexander prevailed over the positive judgments: Droysen turned him into, not 'a crude conqueror', but the soldier, the pioneer, the champion of a superior civilization: Hellenism. That superiority was shown in a flattering light by the depreciation of the culture of the people conquered by the Macedonian:

> While the Empire of the Achaemenids was only an aggregate of countries whose populations would have in common among themselves only the same servitude, there remained in the countries assimilated by Hellenism, even when they had been separated into several kingdoms, the superior unity of civilization, of taste, of custom (Droysen 1833: 696).

Alexander would appear as a sort of king–philosopher, come to rescue the Orient from its barbarism and economic stagnation; Droysen thus repeats to a great extent the theses already set forth in Antiquity, for example by Plutarch.[12] These theses have met considerable success, especially because they were taken up by the historiography of colonialist Europe, which made Alexander into the ideal model of the colonizer: 'We will ask of the Macedonian hero a lesson of colonization which, although more than two thousand years old, is nevertheless, for us, especially today, a burning question of the hour', wrote Commandant Reynaud in 1914. We should not too easily think that this image of Alexander and Hellenism as a superior culture belongs to a bygone era. Alexander fundamentally remains in the opinion of the media the colonial hero of a history made by the 'great men', and despite the considerable progress in the demythification of the history of Alexander in research, Hellenocentrism remains indestructible. We came across this still more recently from H. Bengtson: 'Without Alexander, there would probably not have been world-wide Greek culture; without Hellenism, no *Imperium Romanum*... His life and his achievements

always exists a 'Persian history, fabricated according to the political-ideological preoccupations of the ancient and medieval Jewish community'.

12. *On the Fortune or the Virtue of Alexander* 1.8. See Préaux 1965: 136; Briant 1982: 292-93.

are at the root of many things that exist today' (Bengtson 1968: 329-30). The idea of the superiority of Hellenism is always latent in the present work of some Hellenists, in relation to Transeuphratene.

In the field of Phoenician art for example, we continue to annex to the Greek cultural patrimony the genuine masterpieces discovered in Phoenicia, under the pretext that the Phoenician artists were incapable of being their authors;[13] it has been admitted more recently, however, that the marble anthropoid sarcophagi* were manufactured by Phoenician sculptors, and scholars are beginning to question how well-founded is the attribution of the superb Sidonian architectural sarcophagi* to the Greek patrimony (Elayi 1989b: 257-96). To take an example in another field, it has been thought for a long time that Alexander was at the origin of the transformation of the town planning of Near Eastern towns. The invention of a well ordered plan, with streets cutting at a right angle, could only be, according to this perspective, a Greek invention, and was traditionally attributed to Hippodamos of Milet. Now it is just being discovered that the supposed Greek models are later than the development of the so called 'Hippodamian' plan in certain Near Eastern sites such as Dor.[14]

A third a priori approach is in the process of formation and an attempt must be made to expose it before it develops: Iranocentrism (Kuhrt and Sancisi-Weerdenburg 1987: ix). This deformation seems to have appeared, at least in part, as a defence reaction against the excesses of Hellenocentrism. Faced with a history of the Persian Empire seen through Classical sources alone and exclusively from the Greek point of view, some Iranist scholars have, probably in reaction, shown a certain disdain in regard to the Classical sources and favoured to excess the Achaemenid sources. As a matter of fact, the greatest risk of marginalization of the Achaemenid Empire, assimilated to a despotic and barbarian Orient by some and considered 'peripheral' by others, would be to provoke, in reaction, an underestimation of the historic

13. The theory according to which Phoenician art does not exist goes back to G. Perrot Chipiez (1885: 883), who wrote: 'We can scarcely say that the Phoenicians had any art, in the true sense of the word. . . Like those chemical compounds that are not stable, it decomposes into its elements. . . the only thing that Phoenicia can claim as its own is the formula itself and the title to its blending'(1885: 883).

14. Stern 1990. Likewise at Kerkouan (Tunisia) and at Mount Sirai (Sardinia).

role played by this Empire: let us hope that the new Achaemenid history achieves a balanced evaluation of the continuities and the breaks between the Achaemenid Empire and the Assyro-Babylonian Empire on the one hand and the Hellenistic kingdoms on the other.

Chapter 2

NEW CONDITIONS FOR RESEARCH

The creation, in 1981, of the Achaemenid History Workshops, emanating from a collaboration between the Department of Classical Studies of the University of Groningen and the Department of History of the University College of London, marked a turning point in the history of research on the Persian Achaemenid Empire; regionalist studies of this period must keep this in mind. The results of these Workshops have begun to be published, starting with the third meeting which took place in 1983, in the Netherlands, at the University of Groningen, in a new series appearing at Leiden since 1987 and entitled *Achaemenid History*; the later meetings were held at Groningen or London, but as well at Paris, Bordeaux and Chicago.[1]

The idea behind these Workshops, it seems, was to remove research on the Persian Achaemenid Empire from the Hellenocentric vision to which it appears to have been condemned, without falling at the same time into the trap of Iranocentrism. This effort at objectivity had had, since the 1970s, some precursors, but their efforts had remained isolated (Herzfeld 1968; Dandamaev 1976; Wiesehöfer 1978); the great value of the Workshops was that they made possible the establishment, through the organized meetings, of a real dialogue between Hellenists and Iranists, and the development in this way of a diversified approach to Achaemenid history, using just as easily the Classical sources as the Achaemenid inscriptions. The standpoint, however, always remains that of the central government, even when it is a question of the provinces;

1.　For example: *Le tribut dans l'empire achéménide* (Paris, 1986); *L'or perse et l'histoire grecque* (Bordeaux, 1989). Five volumes have appeared so far: I. *Sources, Structures and Synthesis* (Leiden: E.J. Brill, 1987); II. *The Greek Sources* (1987); III. *Method and Theory* (1988); IV. *Centre and Periphery* (1990); V. *The Roots of the European Tradition* (1990). We have just learned that the Workshops were discontinued after a last meeting in 1990.

thus, the Sixth meeting, which took place at Groningen in 1986, had for its theme 'Local Traditions, Archaeological and Written, in the Various Regions of the Persian Empire'. The explicit objective of this meeting was first of all to get an update on research about the regions, whose extreme diversity and complexity prevent specialists on the Persian Empire from mastering everything; it would be a matter, then, of determining to what extent the latest results of these types of research allow for the clarification of how the central government functioned on the level of the different provinces of the Empire, and to find out whether Iranian society had been affected by its conquests.

A new framework for convergence in historical research on the Persian period has been set up for several years now:[2] it is ASPEP, Association pour la recherche sur la Syrie-Palestine à l'époque perse, whose perspective is basically different from that of the Achaemenid History Workshops; ASPEP is firmly regionalist and centred on Transeuphratene in the Persian period, with any consideration of the central Persian government, or of the bordering regions, only being taken into account in so far as they clarify local problems. The thematic and methodological origin of this new Association is worth keeping in mind.

L'Institut Protestant de Théologie de Paris had created, in 1981, a doctoral research seminar on Studies in Religion, using various methods for approaching documents: historical-critical exegesis, structural anthropology* and historical sociology. Starting from the general theme of 'references to origins', the Seminar has little by little centred its investigation on societies with a scriptural foundation, especially examining the legitimization of power in proto-Judaism and concentrating on finding in the biblical texts the handling of 'key figures' such as Moses or the Patriarchs, at the time when Jewish society, reeling from the shock of its recent history, produced a corpus of traditions soon considered homogeneous and an original reference point. The analysis of the forms of monotheism and of the socio-political conditions for its elaboration in the period of the Second Temple (Persian and Hellenistic periods) quickly made clear the necessity of a precise historical awareness of the societal facts which could have played a role in the structure of the Jewish community (at Jerusalem, in Judaea and in the Diaspora). It would be a matter of uncovering the connection between events in a society or in Scripture and the centralization of cultic activities and the

2. Individual researchers had already been working on their own in this area: for example, E. Stern since 1968 and J. Elayi since 1978.

religious imagination around the Temple of Jerusalem. The Seminar
then became aware of the fact that historical research on Transeuphratene, likely to clarify the origins of Judaism, found itself in a renewal
phase due to recent discoveries, especially in epigraphy and archaeology. It proposed therefore the creation of an independent parallel
group, whose program would concentrate on 'The Societies of Syria-
Palestine and the Near East in the Persian Period'.[3]

The new group began to function in 1985, in an informal way and
in an essentially Parisian setting. As it expanded its contacts, it met a
favourable echo in research circles, both in France and elsewhere, and
had to work hard for new resources, especially through restructuring:
so the Association ASPEP[4] was created in 1988 and it established its
registered office at the Institut Catholique de Paris. As its statutes
clearly state, the ASPEP 'has for its goal to promote research on Syria-
Palestine of the Persian period (sixth to fourth centuries BCE) and to
participate in it, in France and abroad. It intends to encourage interdisciplinary and international exchanges of researchers concerned with
participating in meetings... and in organizing meetings itself... by
taking charge of publications dedicated to research projects on the
region and period in question'.

The first public activities of ASPEP were the organization of an
International Symposium and the starting up of a new specialized
series. The Symposium, which was held at the Institut Protestant de
Théologie de Paris, 29-31 March 1989, with the financial support of
that Institute, the Institut Catholique de Paris and of the CNRS, was
called: 'Syria-Palestine in the Persian Period: Local Governments and
Organization of the territory'.[5] In line with the agreed upon theme,
basic problems, still infrequently studied, could be taken up under the
form of more or less wide-ranging regional assessments (archaeological, epigraphical, linguistic and historical) or in more limited studies:
we asked ourselves, for example, what peoples lived in Syro-Palestine
in the Persian period, what was their life style and what languages they
spoke. How were they governed? What local governments had the
Persians left in place and what was the nature of those governments?
What do we know of the territories where these were put into effect?

3.   Through the initiative of J. Sapin.
4.   Members of the Board: J. Briend, J. Sapin, J. Elayi.
5.   The *Proceedings* of this symposium have been published in *Transeuphratène*
2 and 3 (1990).

In what way was the territorial organization modified by the royal estates and the donations of land? Did Transeuphratene form an administrative unit in the Persian Empire and how can we define it, and so on? These fundamental questions have received partial or comprehensive responses, solidly based or merely outlined; but whether well established or mere possibilities, they give to this new field of research from now on an irreversible factual basis and propose at the same time new ways of posing historical problems. One of the most positive aspects of this first symposium has probably been that it permitted the most varied disciplines to intersect and begin a fruitful dialogue which has every reason to continue.

Conceived in the same spirit as a multidisciplinary series, *Transeuphratène* is at the same time a means of expression to make known the research under way in this field; a meeting place for researchers concerned to exchange ideas and information; and a starting point to direct future research, beginning with reflections in common on the results acquired. It is a series that fills an important gap left by specialized periodicals on the Near East, however numerous and varied they may be. Like the traditional periodicals, *Transeuphratène* accepts research articles relative to its field in all its branches: history, archaeology, epigraphy, biblical studies, historical geography, numismatics, historical sociology, and so on. It contains as well an Information Bulletin on Syria-Palestine in the Persian period with numerous headings, such as: Bibliography, Archaeology, Epigraphy, Numismatics, Old Testament, Book Reviews. Supplied in part by the works of ASPEP, and the meetings organized in national and international settings, that series is the reflection of a local dynamic of research, dedicated to a field too long neglected and one that we wish to see developed in France as well as abroad.

The emergence of a new area of research calls for reflection on the epistemological reasons and conditions that justify it. The creation of ASPEP has been made possible, it seems to us, thanks to the threefold realization of the historical significance of the period, of the region and of the new interdisciplinary openness.

Many researchers who are not specialists on the Achaemenid Empire are today assessing the importance of the period for many disciplines. Thus, Old Testament exegetes pay greater attention than in the past to the process of composition and redaction of the biblical books and especially the two principal collections (the 'Law' and the 'Prophets'), a process which seems to have especially developed in the Persian period. In addition they are seeking, in external documentation, a better knowledge of the living conditions of various segments of the dispersed Jewish community, in their respective contexts and their multiple relationships. The Hellenist scholars, for their part, feel the need to know better the western fringe of the Persian Empire where there took place at that time so many cultural contacts with lasting consequences between the great centres of ancient Near Eastern civilization and the Classical Greek world; this need is more especially felt by specialists on the Hellenistic kingdoms who are unceasingly and more and more sent back to the Persian period in connection with the problems that they meet. Finally, archaeologists who specialize in the Middle East of the first millennium are becoming aware that the Persian period, which long remained almost imperceptible, often has significant surprises in store, both in excavations and in surveys and even in chance discoveries, to the extent that they better discern the different components of these material aspects.

It seems too that scholars have become aware of the advantage that can flow from the study of the human characteristics of Transeuphratene in the setting of the Persian Empire. We have known, certainly for a long time, the advantage of its geographical position which made it a regional crossroads, but we are today especially interested in its diverse ethno-cultural components. Biblical scholars, aware of the recent epigraphical discoveries, especially in Palestine, and sensitive to their onomastic and linguistic contribution, are now giving greater attention to the people of the region; to the spatial distribution of the various ethnic groups; to the districts of mixed population; to the socio-political, economic and cultural concerns that motivated them; to the organization of the area and its administration by the Persian chancellery, and so on—all this in order to understand better the regional context of the Jewish community. Hellenist scholars, for their part, taking note in the Greek cities of the effects of intercultural* relations with the Phoenician world, look for their causes (starting points, groups in contact, material or other supports, trade routes, economic and political conditions), which implies a precise knowledge of the whole Levantine coast region and its hinterland; always very interested in Greek expansion, they also look there for every trace indicative of a Greek presence. The archaeologists feel still more involved when it becomes time to map out the different collections of objects from the

Persian period, collections still very incomplete in certain parts of Syria-Palestine, and yet indispensable in the areas where there is a lack of the Greek imports that have served until now as the principal indicator. More than ever, it is to the mosaic of local cultures scattered in the different areas of Transeuphratene that the archaeologist of the first millennium has to be attentive in order to make evident the specificity of that regional corpus, with its new traditions and trends, flowing from other centres of civilization and more or less integrated into original syntheses.

In the third place, all the researchers who work on Transeuphratene, no matter what their specialty, are faced with a documentation that is so incomplete, diversified and dispersed, that they can with difficulty get away from becoming involved in multidisciplinary and interdisciplinary research. That necessity imposes itself, it seems to us, in an inescapable and conscious way, but all its practical and methodological implications have not yet been grasped.

As a matter of fact, the opening of a new field of research, especially when it is as extensive and full of pitfalls as this, necessitates an experimental phase at first. On the one hand, we do not think that we can put into practice right away, in a mechanical and artificial way, methods of approaching it that have proved useful elsewhere. On the other hand, the first results that have been obtained represent a necessary step, but one meant to be surpassed, which is true in a general way, but more especially at this stage of methodological gestation.

In an area in which researchers with such diverse specialties and interests begin to meet, it seems necessary to define at the outset a minimum of common objectives: let us say, perhaps, the knowledge as objective as possible of a region at a given time. Such a proposition would probably be enough to earn for us the label 'positivists',* to which we can reply: 'If we are not yet aware that we are perishing from divisions, fractions, compartments, when will we know it?'[6] We are fully aware that the era of positivism is long past and that any analytical work is a function of a reference system, and reflects the subjectivity of the author and the temporal context to which one is

6. Febvre 1954b: 208. This chapter is an excerpt from a paper presented by J. Elayi to the 1990 International Meeting of the SBL, held in Vienna, 5–8 August 1990: 'A Reflection on the Place of History in Research on Achaemenid Transeuphrates' (published in *Trans* 4 [1991]).

attached;[7] this, by the way, applies not only to history, but to the social sciences in general: the 'falsification' of documents in epigraphical works or in textual criticism, for example, is not unknown in our field. Nevertheless, it seems possible to ensure a minimum of objectivity by rejecting any intentional 'falsifications', of any kind whatsoever. We no longer think of research on Transeuphratene in the Persian period as random undertakings to collect 'facts' or 'documents', whose meaning will perhaps become self-evident later: it would be absurd to begin in this way the exploration of the vast areas of obscurity that await us. More particularly in our field, we have need of some main themes which as we progress we will be able to drop or replace, whether it be directions, orientations, working hypotheses usable in a transitory way and as models, on condition that they be relevant.[8]

The disciplines, old or new, constitute our access routes to the new research area; their practice cannot depend on traditions or fads, but on what has proved through use to be effective. Some traditional disciplines always give excellent results, others need adaptation, others are out of date or useless. A similar sorting out ought to be done for the new disciplines, taking into account the further fact that several among them are still in their infancy. Thus, historical sociology can be useful, on condition that it is really adapted, as we will see in Chapter 10. Econometrics*, which is situated at the crossroad of statistics and economics, cannot be of use to us at the moment, because of insufficient data. Serial history, in the strict sense, which studies cyclic crises, is of limited use in our area where the data are very intermittent in time (Furet 1974; Chaunu 1978), but the concept of seriality is valuable, especially in epigraphy where the inscriptions are 'put in series' to clarify one another.

All the disciplines which, with use, are found to be helpful in our area, even the most specialized, have need today of an interdisciplinary setting to be adequately practised, as we have already emphasized. The interest in interdisciplinarity is obvious: '... research, in the area of the social sciences', as P. Chaunu, for example, wrote in 1978, '[is] each

7.   See, for example Popper 1934 (on the 'principle of falsification'); Aron 1938: 350-55; de Certeau 1974: 5-7.

8.   Febvre, 1946: 7, wrote: 'I am asking them (the historians) to get down to work as Claude Bernard would, with a good hypothesis in mind. Never to make themselves collectors of facts, in a haphazard way, as one once sought out books on the quays. To give us, not a mechanical history, but a problematic one.'

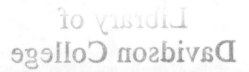

day more interdisciplinary, because it is on the frontiers, to be sure, in the shock wave of unusual parallels that burst forth, fructify, the innovation, hence the discovery' (Chaunu 1978: 28). But what kind of interdisciplinarity? Even if an all-embracing vision is not claimed for each one's own discipline, the specialist can no longer be content with a modest practice of the specialty, while closed up in an ivory tower.[9] As Braudel rightly stated:

> Nothing would be better, of course, if those engaging in a legitimate specialization, laboriously tending to their own gardens, are forced nevertheless to keep up with the work in a neighbouring discipline. But the walls are so high that very often they block the view (Braudel 1969: 33).

In fact, it is not just a matter of being open to different disciplines on their fringes, where the reflection on many different levels runs the risk of a gratuitous game of cross questions, ill-founded and of no consequence. We no longer really see any usefulness in quarrels on the boundaries of competence that separate each discipline from its neighbour, even if each legitimately tries to define itself in order to locate and preserve its place in the ceaselessly growing collection of social sciences. Specialized research can only assume its true scientific dimension if it is contextual, that is, if it is inserted into a comprehensive programme, with precise objectives from the beginning. In the United States for example, interdisciplinary activity has taken the form of collective research by groups of specialists, coming from different backgrounds in specific cultural areas ('area studies'). But the setting up of such groups does not necessarily imply their success, since the elaboration of a good common programme is difficult, just as is the route from the juxtaposition of competencies to dialogue and exchange. From this there follows the importance of a real dynamic for research developed in an environment adapted to exchanges.

It is here that multidisciplinarity comes in and where the role of the historian can be situated in our field. One cannot be content, in fact, with juxtaposing the results of various disciplines, results which are always very relative, depending on the documentation which provided the foundation and on the method of analysis and elaboration which

9   Garelli (1964: 126) had already observed in regard to Assyriology: 'Basically, what most damages Assyriology is the anarchical dispersal of efforts. A lot of isolated research efforts should be coordinated.'

guided it. The researcher must be initiated into the disciplines indispensable for carrying out the research, while having confidence in the
conclusions of specialists in the other disciplines: it is a difficult task,
to be sure (all the more so since certain disciplines are themselves
multidisciplinary), and one calling for continual keeping up, but not
impossible, if intelligently conceived. We are really far from the time
when an encyclopaedic knowledge was perhaps conceivable and we no
longer really trust the most recent claims about a total history: multidisciplinary history should, in our view, limit its ambitions to a
restricted spatio-temporal area and to relatively simple and circum-
scribed related problems; we will distinguish, of course, this history
'at ground level' from histories at other levels, carrying out more or
less wide-ranging studies, while utilizing secondary sources or even
syntheses of secondary sources. These other forms of history, which
have their usefulness too (Chaunu 1978: 119), have scarcely any place
for the moment in our new field of research, where it would be altogether premature to go beyond the stage of partial and provisional syntheses. Multidisciplinary history is possible, then, if it has limited objectives and if, on the other hand, the historian does not claim to compete
with experts in specialties to which they have exclusively devoted themselves: it is only a matter of having sufficient initiation into the specialties necessary for understanding and commenting on the work of
specialists, in order to set up a fruitful dialogue with them, and to
resolve, should the occasion arise, some specific point important from
a historian's point of view, but which has not yet attracted the specialists' attention. As J. Le Goff and P. Nora acknowledged in 1974, it
is true that the role of the modern historian is an uncomfortable one:

> More and more specialized, the historian has not however attained a tech-
> nical expertise which, on the one hand, would distance him from asso-
> ciation with second-rate popularizers, the scribblers of anecdotes, and, on
> the other, put him on a level with the new scientific heroes of the latter
> half of the twentieth century . . . (Le Goff and Nora 1974: xiii)

Nevertheless, the multidisciplinary initiative of the historian, all the
more detailed as his area of research is limited, allows him to domi-
nate better than the specialists, even those most open to the work of
neighbouring disciplines, because he can integrate all the necessary disciplines into his initial term of reference, correcting and completing it
eventually. As well he is naturally led to play a role on the level of the
elaboration of programmes and of the coordination of research: 'the

collection and sifting of facts, the arrangement of index cards, all this work of quarrying is only worthwhile if one uses it as an architect would', wrote L. Febvre (1954c: 109). It would be wrong to consider this vision of history as 'imperialist'[10] since we think that the historian always has much to learn from specialists, despite his multidisciplinary initiation; and, on the other hand, a broad openness should precisely make it possible to promote real interdisciplinarity, capable of putting an end to the 'imperialism' of certain disciplines. All the disciplines could enjoy the same epistemological status (a critical appreciation of the value and significance of every document in a logic of historical knowledge), and they could mutually challenge one another without worrying about precedence or hierarchy, in a free openness providing an opportunity for a full flowering of methods and for their mutual enrichment. In the end, the discipline of history, by its essence integrating, leads to teamwork.

All the syntheses that can be envisaged on Transeuphratene in the Persian period cannot be carried out at the same time and are even far from being able to be done, since extensive geographical sectors of this region are still devoid of the least published documentation. But, if a start is made from the documentation accessible today and in the process of being brought together by way of very useful appraisals, some new problems of all kinds would come up, drawing attention to regions little explored and to fields neglected up to the present. Interdisciplinary openness well understood is not a fad, but is becoming a necessity: it is already indispensable for some researchers who are trying it and will become logically unavoidable when research programmes integrating a broad range of disciplines are put in place. These programmes must have clearly defined scientific objectives, be well targeted thanks to historians, precise thanks to specialists, and open thanks to a common elaboration by all those interested, and must have sufficient means provided through a policy of intelligent and responsible research. At present, the decompartmentalizing of disciplines has begun to take place in the dialogue which has been established between researchers taking a close or remote interest in Transeuphratene of the

10. This is the reproach addressed, among others, to Braudel: Barraclough 1980: 378.

Persian period, letting it augur well for the future. We intend to study in the following chapters some examples of the adaptation of old and new disciplines, under certain conditions, and to show how their decompartmentalizing can be carried out.

Chapter 3

THE USE OF COMPUTERS

Contrary to what we might think, the social sciences made use of data
processing very early; its introduction into studies of the ancient Near
East originated in the research of J.C. Gardin in 1955–56 at Beirut,
supported by H. Seyrig, director at that time of the Institut français
d'archéologie orientale. Being especially interested in the problems of
method, Gardin worked out a certain number of classification codes
for very diverse Near Eastern objects: vases, Bronze Age tools, Mesopotamian cylinder seals, coins, Koranic concepts.[1] Some users did take
advantage of the important analytical work which was being carried
out at that time, but Gardin's initiatives were nevertheless seldom followed, perhaps because he had not from the beginning advocated recourse to computers, with which the French were still not yet familiar
at that time (Cleuziou and Demoule 1980: 7-15). Shortly after this first
attempt, in the course of the 1960s, a similar movement developed in
the United States, initiated by L. Binford, who used the computer right
away, since his compatriots were already quite used to it (Binford and
Binford 1966; see Hours 1980: 9-10, 18); as a result, that advance was
in this case followed. But it is especially since the 1970s that the progress in thinking and the improvement in the machines favoured the
attempts at adaptation of the computer to the social sciences, especially
to archaeology, and so the projects multiplied. Most of the researchers,
however, remained opposed to its usage, or were content with simple
deductions and percentages; it was only adopted by some of those who
frequently provided their works with lists in an appendix, but its usefulness was incidentally not always evident. The relationship that researchers maintained with the computer was not the same as it is today.
Computing work was mostly carried out in a centralized way, on
machines that were relatively powerful for that period; they were set

1.   J.C. Gardin 1955; 1958; cf. on this subject F. Hours 1980: 9-10, 18-19.

up in teams around some better trained or more motivated researchers, or around a data processor, for those who had the luck to be able to come across one. It was the time in which there was enthusiasm about computer applications in the different branches of the social sciences, and where they took advantage, in the scientific community, of an advanced methodological attitude.

The computer is today no longer the prerogative of a few passionately interested in a sophisticated methodology, or of initiates in quantitative and formal disciplines. Far from being reserved to a few specialists, it can now be within reach of all researchers. If one can always do good research without recourse to a computer, one can no longer deny its considerable contribution to the social sciences (Auda 1987). The arrival of personal computers marks a new stage in computing with light equipment, making possible wide decentralization. But their wide diffusion in research poses specific problems since we do not yet have a level of maturity in this field. A first series of problems arises from the quantity of disposable information, the cost of its acquisition and its processsing. In our field, it is documentary research that is called for in the first place: the automatic documentation by computer permits the construction of powerful documentary systems, namely, the setting up of large collections (data banks), organized according to particular classification systems (databases) to be accessed according to selective criteria. But it quickly became apparent that the creation of large information data banks had a prohibitive cost, especially because of the regular regeneration necessitated by the conservation of the documents, and because of the fragility of the supports and the rapid evolution of the machines; such banks cannot be conceived except under the form of enterprises devised on an international scale. The optimizing of documentary research perhaps takes place better through computing networks that rely on less ambitious, more specialized and more numerous banks, with the utilizing of these networks being supported by on-line data processing.

A second series of problems, resulting from the decentralizing of resources, arises from the fact that each researcher attempts to set up informational data banks on the object of that individual's research. The microcomputer makes possible the assembling of all the data on which the researcher intends to work, whether it be a matter of texts, or of archaeological, epigraphical, bibliographical, or other data. This mass phenomenon could lead to an enhanced value for the information

owing to the setting up of complementary data files, their multiplication and their circulation. But it is nothing of the sort, since the microcomputer encourages individualistic attitudes on research: each one
undertakes the creation of a personal base in a private way, which
leads to a loss of time since there is the possibility of the work having
already been done. It also puts a brake on research since efforts are
scattered and anarchical. Even when there is a willingness to share computer data, the multiplicity of the materials and of the methods of gestation can be an obstacle. It is always necessary, therefore, to have prior
agreement by the researchers wanting to create a database, and to have
sufficient financial support in order to procure compatible equipment.

There is a tendency to forget sometimes that the computer remains
a tool and that its use presupposes competence and a critical attitude.
The descriptive aspect has often taken precedence over careful reflection: there has been a proliferation of computer treatments of various
collections, whether texts or objects, carried out with heavy invest-
ments in money, personnel and time, and the benefits produced from
them are really meagre. It is especially the private use of microcom-
puters that encourages the thoughtless attitude which amounts to say-
ing: 'Let us first create our database and we will then see what we will
be able to do with it'. It is understandable that as a first stage, the
descriptive tool offered by data processing would so fascinate researchers that they would forget to formulate their hypotheses in
connection with it: if this experimental stage was necessary, it is clear
that today we have to go beyond it.

What can we reasonably expect from computing in research on
Transeuphratene in the Persian period, starting from some of the
knowledge already acquired and from the advances noted in the other
fields of the social sciences? Textual material has given rise to many
computerization projects, especially centred around the manuscripts of
the Bible.[2] According to some (Claassen 1988: 292-93), recourse to
computers in the biblical field had its origin in the deficiency of most
of the traditional approaches and in the feeling of not being able to get
control of the mass of documents, which R. Rosenthal called the 'crisis
of Orientalism' (Rosenthal 1983). The computerized processing of biblical texts began to develop a quarter of a century ago, with a number
of pioneering centres, notably the first publications of the *Computer
Bible* by the American J.A. Baird of Wooster, the works of G.E. Weil

2. Cf. Homan 1988, for a state of the question.

at Nancy and those of A.Q. Morton at Edinburgh (Bajard and Servais 1986: 9; Bajard 1986). Since then many other centres have started up, such as the Centre d'Analyse et du Traitement automatique de la Bible (CATAB) at the University of Lyon III; the Centre Informatique et Bible at the Abbey of Maredsous, in Belgium; the Department of Near Eastern Studies of the University of Michigan and the Computer Workshop of the Theology Faculty at the Free University of Amsterdam.[3] The research has developed into several specializations, with the three main ones being concerned with the language of the texts, textual criticism and literary and exegetical questions.

The first area of specialization has been by far the most fruitful. Those involved have applied themselves to creating lists of words and concordances*, which provide the basic tools for biblical research. Thus, more than 20 volumes have already been produced by the team of J.A. Baird and D.N. Freedman at Wooster, in the American series *Computer Bible*, already mentioned, which has just completed the series begun in 1983, *Instrumenta Biblica*, published at Amsterdam (Postma *et al.* [eds] 1983; Abercrombie 1984). The creation of more elaborate tools has also been undertaken through the selection of lists according to certain criteria, lexical or semantic for example (Claassen 1988: 289 n.17). The use of these various lists and concordances has given rise to numerous studies, monographs and articles, as, for example, that of E. Talstra on the use of the particle *kēn* in biblical Hebrew (Talstra 1981). A special effort has been made recently, especially at the South African University of Stellenbosch, to develop syntactic databases, not only in biblical Hebrew, but also in the other ancient semitic languages.[4] The importance of this project is that it makes it possible, for example, to record the virtual relative clauses which are not indicated by the use of a relative pronoun and which would not therefore appear in the usual listing processes.

The second area of specialization is much less fruitful, but still promising. For Weil, the computer provides the only means of preserving the rich variety of some aspects of the Massora*, that is, the traditional philological commentary accompanying the Hebrew text.

3.   We must mention also, among others, the Department of Semitic Languages of the University of Stellenboch and the Old Testament Department of the University of Tübingen.
4.   Claassen 1987: 11-21; Bothma 1989; likewise at Palo Alto in California; Forbes 1987.

The database CATAB, which he has designed, contains all the variant readings and the paleographical information for the Hebrew manuscripts and their first translations (Weil 1964–65; 1986). The University of Pennsylvania and the Hebrew University of Jerusalem are working on a textual criticism project on the Septuagint, and the Abbey of Maredsous, in Belgium, is preparing a *Concordantia Polyglotta*, that is to say, an index of the Bible in its different translations, at the same time analytical and multilingual (Kraft and Tov 1981; Vervenne 1981). It must be kept in mind that the number of biblical manuscripts amounts to almost 30,000, with all the languages joined together, without counting several million patristic* citations. The solution which has been envisaged consists of breaking up the difficulty into its component parts and collecting together the data concerning the manuscripts in just one place where the data would be available for consultation from a distance: this is the long-term project of the data bank of the Centre de documentation sur les manuscrits de la Bible, at Montpellier, which works in collaboration with l'Institut de recherche et d'histoire des textes, at Paris and Orléans.[5]

The third area of specialization is that which is concerned with literary and exegetical questions. The analysis of the form and that of the content of a text follow very different procedures: while the first is similar to a literary approach and comes under discourse analysis, the second is much closer to the logic of the experimental sciences. Now, some researchers have tried to pass directly from quantitative linguitic data to literary matters, and have thus made 'discoveries' which have sometimes engrossed the media, but which in general have not convinced the exegetes (Claassen 1988: 268-87 and nn. 5-6). This was the way that Y.T. Radday wanted to demonstrate the unity of the book of Isaiah solely on the basis of statistical information obtained through computer analysis, and called into question the 'documentary hypothesis'* in regard to the book of Genesis (Radday 1973; Radday, *et al.* 1982: 481). The pitfall to be avoided in this field is to think that the computer can be a substitute for the traditional approaches: philological, linguistic and exegetical.

All in all, the application of computer methods to the Bible has fallen short of the hopes that had been placed in it. Many projects, sometimes very alluring, have not been carried out or, when they have

5.   Firmin 1986: 402-403; this centre has become the GDR 797 ('Histoire du texte de la Bible').

been, have not yielded the expected results; nevertheless, we must be aware that some projects cannot produce immediate results, but can shift the orientation of research in the long term. What appears very clearly today is the need for better communication among all the researchers concerned, in order to exchange methods and results: in fact, information on the research projects under way circulates poorly; some local projects are completely ignored and quickly get out of date because of lack of openness to the international scientific community; finally, the publication of works of this type are not always accessible, since it is done quite frequently elsewhere than in the specialized biblical reviews (Claassen 1988: 284-86). However, despite all the criticisms that can be raised about the application of computing methods to the Bible, the results already obtained are such that it could be wished that research on the other North-West Semitic corpora would make up for its delay on this level, and that the researchers would go beyond the phase of individual databases, all the more jealously guarded since the corpora are small.[6]

For some years, the use of the computer in archaeology has been consolidated and intensified: data processing just completes the inventory of the data. Archaeology today is no longer content to list objects: it needs a tool that utilizes the means of analysing data just as easily as it does their spatial representation. Computing in archaeology has served first and foremost to set up bibliographical and documentary data banks permitting access to a general or a specialized background;[7] for example, we may cite FRANTIQ, a database resulting from the fusion of catalogues of several libraries, which includes the archaeology and history of the ancient Near East. Furthermore, microcomputing permits the easy formation of data files of information from the excavations or the laboratory. But there still remains the problem of learning to make a good choice of the data and the means of processing, and especially to define the objectives of the computerization.

The first archaeological expedition to the Near East that elaborated a data bank was that of R.J. Bull, at Caesarea (Straton's Tower),

6.   We note, however, that computerization has begun to be used in Ugaritic and Amorite studies: see, for example, Whitaker 1972; Bordreuil and Pardee 1989; Gelb 1980. It is currently practised in Assyriology (for example, *Annual Review of the Royal Inscriptions of Mesopotamia Project*, I, 1983).

7.   Tunca 1987; Ryan 1988: 3-27; F. Djindjian, *Méthodes pour l'archéologie* (Paris: Armand Colin, 1991).

beginning in 1970 (Strange 1984; Toombs and Wagner 1971): this site had produced an enormous quantity of material from the Roman and Byzantine periods (pottery, coins, sculpture, lamps, etc.) that could not be registered by traditional means; the other pioneering expedition in this field was that of E.M. Meyers at Khirbet Shema' in 1971–72 (Meyers *et al.* 1976: 269-80). The second step was the utilization of a terminal right in the field in the excavations at Meiron in 1977 (Strange 1981); the advantage was to have immediate access to the database in order to keep control of it and be able to take adequate initiatives in the course of the excavations. Successive more or less steady improvements in the system of recording data were introduced by Z. Yeivin of the Israel Department of Antiquities, then by T. McClelland of the University of Pennsylvania, who worked at developing a uniform classification system of all published Iron Age pottery; this allowed him to develop five significant regional occupational sequences: that of Hazor; that of Deir 'Alla; that of Lachish and Tell Beit Mirsim; that of Megiddo, 'Amal and Beth Shean; and that of Tell Jemmeh, Zuwayid and Tell el-Far'ah (McClelland 1979). The first complete database of information about terrain in Near Eastern archaeology had been that of the Heshbon Excavation Project at Heshbon in Jordan, directed by L. Geraty and beginning in the 1980s. 'System 2000', developed by Sheila McNally and Vicki Walsh for the material from Tell Akhmim in Egypt, again marked a new step, since it permitted the classification of each new object by making comparisons with objects from the same cultural area or from other cultural areas (Strange 1984: 133-35 [with bibliography]). A new system for the management of data banks was tested out in 1983 in the excavations at Sepphoris in lower Galilee, by Oklahoma State University (Chenhall 1981:2): this system allows for the correction of old information, the integration of new information and the deletion of what has become inadequate.

The analysis of data constitutes the principal contribution of the computer to archaeology (Strange 1984: 138-42 and nn. 28-44). Spatial representation, with selective retrieval, allows for the visualization of global or individual situations; the analysis can be made either on the microlevel, that of the site, in order to determine the areas of specialized activities, or on the macrolevel, that of the region, to demarcate the zones of cultural similarity. More sophisticated methods, based on artificial intelligence, such as the use of expert systems*, allows for a measurement of the extent of the area of application (Gardin *et al.*

1987; Lagrange 1989). Displaying of temporal structures, or seriation, had already attracted the attention of the very earliest archaeologists, such as F. Petrie in 1900: it is a matter of dealing with certain types of objects, which are impossible to situate either in an absolute chronology or even in a relative chronology (surface deposits without connection to any stratum, illegible strata, necropolises), but whose presence can each be recognized continuously through a period.[8] Factor analysis has made possible the simultaneous representation of types of objects and levels, thus leading to the identification of cultural features and to their classification through the most rare or the most frequent types of objects over a short period.

Simulation*, a hypothetical-deductive approach, has been used by American 'New Archaeology' since the 1960s, in the field of social and cultural anthropology, especially in the prehistoric area; then it penetrated into Europe, where it was applied to problems that were more historical (see, for example, Sabloff [ed.] 1981). So we have gone from research on correlations between the variations in the material aspects of cultures and the modifications of the ecological and cultural context (eating habits, technological advances, social organization, strategies for the usage of space) to research on the causalities and the process in the phenomena of acculturation*, invasion or trade. Because of its multidisciplinary openness, simulation in archaeology has extended the field of interpretative hypotheses, but has shown itself to be open to criticism because of its hasty deductions: experimentation which, in the field of the exact sciences, allows for the adjustment of the mathematical models according to the phenomena observed, is lacking here. The value of the simulation always depends on the abundance of information, on the reliability of its possible valuation controlled in socio-economic terms, and on the exploitation of numerous and well conserved deposits; important progress still remains to be realized in this field.

The methods perfected by the geographers and ecologists in spatial analysis are likewise the concern of simulation. Their application to archaeology comes up against the problem of different conditions in the discovery or the conservation of small sites or deposits, of which we can draw only very rarely representative maps: the methods of spatial sampling and the cartographic scales used must be tested and compared so that a relative or absolute coefficient of representativeness can be

---

8.   Petrie 1900–1901; on recent research in this field, cf. Djindjian 1985.

applied to them; the imprecise limits of the sites also necessitates weighing according to their forms and their spatial contexts so as not to invalidate, for example, the Nearest Neighbour analysis. On the other hand, the methods of regression can be useful: they have made apparent, among other things, the effects of the mode of transport by land or by sea on the distribution of pottery. Finally, it turns out that the theory of central places*, applied to historical (and protohistorical) archaeology, remains valuable if we have to deal with a society based on a traditional agricultural economy and a relatively solid local socio-political structuring.

The applications of data analysis are still not developed much in our field, especially in the case of the most sophisticated ones. We may point out the study by J. Raynor of the distribution of coins at Nabratein and in the neighbouring sites, using statistical methods, which, however, lacked precision (Meyers *et al.* 1981). Since 1978, A. Zertal has been directing a research programme in the region of Samaria, consisting of an exhaustive regional survey and the excavation of key sites. This programme uses data processing to store the data, to establish typologies and to make maps of the results; the first general survey, recently published (Zertal 1990), seems very promising with regard to the demographic and economic development of Samaria in the Persian period, but we must wait for the publication of the computerized procedure used to be able to assess its validity.

The history of Transeuphratene in the Persian period will be the last example which will serve to illustrate the use of computing methods, but it is possible to find many others. Historians have used these methods for a good quarter of a century, which has intensified and facilitated the directions of research, but has not provoked epistemological breakdowns. The great turning point in the discipline of history had been tackled and settled well before the introduction of computing in the social sciences in the years 1930 to 1950. However, the use of the computer changed the historian's methods of work insofar as there was a transformation in the approach to the documentation, whatever kind it might be, quantitative or qualitative.[9] It obliges one to utilize, define and clarify the sources to be used in the investigation, a task that would perhaps have been avoided, if it were not for the constraints of the

9.   Zysberg 1986; see also Djindjian 1991.

machine; the computer gives access to a better handling of the information and perhaps makes possible the envisaging of problems that would have otherwise been impossible. In the case of sources requiring exhaustive perusal, work is done more and more today by way of sampling.

The databases developed by the historians are more often heuristic than documentary, that is to say, conceived not to absorb all the available information in a given field, but to allow the handling of the collected information in view of resolving a specific problem set out beforehand. A heuristic database can easily be reactualized, which allows for the verification of whether the results of the analysis of the data are to be modified; on the other hand, it can be broadened by the addition of new parameters, if one later on wishes to take up other problems. As an example of an application in our research field, we can cite the study of Greek imports into Phoenicia in the Persian period (Elayi 1988: 207-12). The general problematic in the work which was intended to measure the penetration of Hellenism into Phoenicia before the arrival of Alexander called for the consideration, among various sources of information, of several thousands of archaeological documents, namely, imported Greek products or local products marked by Greek art. It was necessary to classify rapidly these documents according to the criteria selected, in order to carry out the necessary selections for the requirements of the study and to verify various working hypotheses. Unlike documentary data banks (objects from excavations or museum collections, for example), whose programmes endeavour to take into account all the characteristics of objects, it was not necessary in this study to take all of them into account, even when they were known. As for objects presented in an incomplete manner in the excavation reports, they could provide useful information and had to be taken into account too: it sufficed to introduce some parameters indicating whether each characteristic of the object was certain or not, or unknown. The programme was set up in a way that was commensurate with the accuracy of the excavation reports and of the requirements of a specific historical problem, while omitting the non-significant categories of objects and their characteristics.

These few examples have been put forward to show all the profit that can be drawn from the use of the computer in different disciplines useful for our new field of research, provided there is the knowledge of how to handle it competently, in a critical spirit and prudently.

Must specialists on Transeuphratene for that reason become computer scientists, a new category of what J. Raben and S.K. Burton have called 'hybrid scholars'?[10] At the present stage of the evolution of computer science, to be a computer scientist is a full-time occupation; even with the help of microcomputing, the processes of encoding and keyboarding of data often remain difficult, long and tedious. One of the most desirable solutions would perhaps be, in the present transitory stage, close collaboration between antiquity specialists and computer scientists, with a minimum of initiation of each one in the methodological approach and the problematics of the other.

10. Raben and Burton 1981: 248; Claassen 1987: 19 has confidence in this kind of researcher and issues an appeal for collaboration.

Chapter 4

TOWARDS ANOTHER ARCHAEOLOGY

Archaeology was born with the Italian Renaissance, drawing on the sources of Western civilization, Graeco-Roman and Christian. In the enriched Italian cities and in the Papal State which remodelled Rome, architects and artists revived the arts of Greek and Roman antiquity, rediscovered in the texts which they were going to call 'classical': everywhere surveys of monuments, designs of sculptures, reliefs and inscriptions, collections of gems and coins multiplied. These new practices, joined to a new notion of the universe that put humans and their most brilliant works at the centre of all understanding and all power, were not a simple passing fad then, but the manifestation of an ever more fixed and well thought out curiosity, which became little by little a science, but one that already, quite rightly, could be called 'archaeology'.

A few European pilgrims set off on the route to Palestine a good century after the end of the Crusades, and became more and more numerous, beginning in the sixteenth century, with Palestine from then on being under the Ottoman authority.[1] They seemed to pay more attention to what they had discovered than to their own pious acts, systematically observing, taking notes, sketching the monuments (and not only those of the Christian holy places), the inscriptions, the landscapes, the sites, the plants, the native populations, in order to enrich European Christian civilization from one of the sources that it claimed as its own. The reports of these voyages illustrated in this way grew in number in the sixteenth and seventeenth centuries, so much so that a Dutchman, A. Reland, rather than travelling to Palestine, preferred to recover the ancient and modern documentation, which had become so abundant, in order to present it in a critical manner, gathering it together in a learned

---

1. See the account of the 'discovery of ancient Palestine' in Albright 1962: 23-48.

monumental survey which he published in 1709: *Palaestina ex monu-mentis veteribus illustrata.* In 1737, Bishop R. Pococke made a journey through Syria-Palestine off the beaten track, from which he brought back a rich harvest of plans, sketches and copies of inscriptions, largely unedited. In the middle of the eighteenth century, right in the middle of the age of the 'Enlightenment', the return to the style of Classical Antiquity increased still more the stream of voyages. It was especially the Orient, near and far, which attracted and became the object of a new knowledge that was at the same time exotic and self-justifying for the Occident, so sure of its superiority. At that time there began the scientific voyages and missions to Pharaonic Egypt, Avestan Iran and Hindu and Vedic India, and to the more familiar impressive Roman remains in Syria, which had already filled the preceding travellers with wonder. In this way there were carried out, for example, the surveys of the ruins of Palmyra and Baalbek by R. Wood and J. Dawkins, published jointly at London and Paris in 1753 and 1757.[2]

During all these centuries of awakening, archaeology was limited to establishing museographic, architectural and epigraphical collections, according to criteria that were first and foremost aesthetic, and had required practically no excavating. Its ground investigations consisted above all of surface explorations, but always more systematic and more specialized: at the end of the eighteenth century, Bonaparte's expedition to Egypt, with his group of scholars of all disciplines, was its most striking illustration. That form of archaeology had the great merit of preserving for the memory of humanity the representation of monuments, objects and inscriptions which have disappeared since then. It is more than ever indispensable today, in the face of the frenzy of modern excavation work and the proliferation of clandestine excavations. In the countries with ancient civilizations, the inventory of sites (never finished), their protection and the mounting of salvage excavations remain the major preoccupation of all the national Antiquity services. Consequently, modern archaeology, more conservative than ever in the good sense of that term, cannot repudiate, without repudiating itself, the work of documentary collection of a nascent archaeology, imperfect as it might be.

2.   Dussaud 1931: 5-22 is a handy introduction to the first archaeological activity in Syria.

At the end of the first age of archaeology, the explorations led by genuine scholars in the course of the first half of the nineteenth century led to a decisive step being taken in surface exploration: in the Near East, U.J. Seetzen, J.L. Burckhardt, C.L. Irby and J. Mangles discovered certain leading sites in Transjordan (Amman, Jerash, Petra, Iraq el-Amir); T. Tobler and especially E. Robinson and E. Smith produced a revolution in the historical geography of Palestine by identifying numerous ancient place names (Albright 1962: 25).

About 1850, archaeology of the Near East passed through a new stage, by taking on the clearing of monuments: thus, F. de Saulcy cleared north-east of Jerusalem the hypogeum tomb of Helena of Adiabene and her sons, which he took to be that of the kings of Judah (de Bry 1982: 56-58). It is in this context of a 'turning point' in archaeology that the mission in Phoenicia directed by Renan took place. The chance discovery, in 1855, of the sarcophagus of 'Eshmun'azor, king of Sidon, had aroused among European scholars the hope of new discoveries, to be made if possible in better conditions of scientific control. More than any other, Renan was already ready: in his 'Mémoire sur l'origine et le caractère véritable de *l'histoire phénicienne* qui porte le nom de Sanchoniaton traduite par Philon de Byblos' ('Record of the origin and true character of the *Phoenician History* which bears the name of Sanchoniaton translated by Philo of Byblos'), he had painted a masterful picture of the Hellenization of Western Asia and emphasized the interest that there would be to excavate the soil of ancient Phoenicia, in particular at Jebeil (Byblos). The occasion presented itself in 1860 with the intervention of the French troops in Lebanon: in imitation of his uncle, Napoleon III organized, with the army as a logistical base, an archaeological mission which he entrusted to Renan. Thus, four sites were opened, of which three were supervised by Renan himself, at Amrit, Jebeil and Tyre, and the fourth by Dr Girardot at Saida (Sidon); a fifth at Oumm el-'Amed (Hamon) was entrusted later to the architect M. Thobois. Renan prospected the country as well, to discover other monuments and inscriptions. The success of his excavations and his survey was the establishment, through archaeology, of Phoenician history, of which he was the founder: 'History, he had written, is the fruit of the immediate study of monuments... but the monuments are not accessible without the research of the philologist and the archaeologist'.[3]

---

3. Discourse at the Academy, 29 December, 1871.

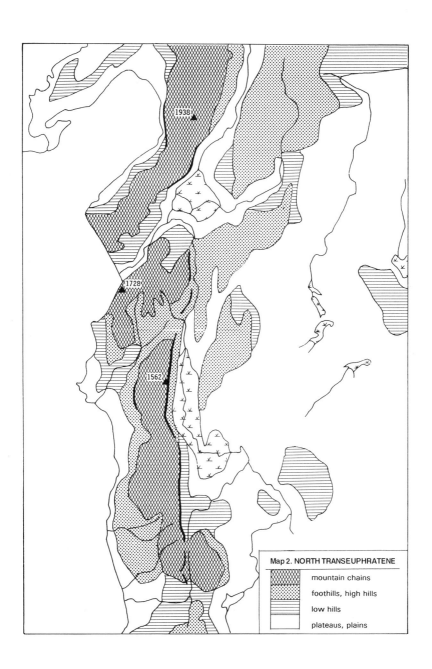

Map 2. NORTH TRANSEUPHRATENE

mountain chains

foothills, high hills

low hills

plateaus, plains

1938

1728

1562

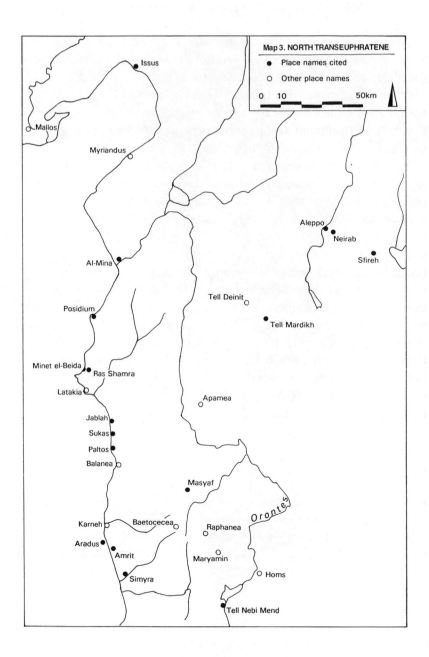

Map 3. NORTH TRANSEUPHRATENE
● Place names cited
○ Other place names
0   10   50km

The publication of his *Mission de Phénicie* in 1864 opened new horizons for the Western scholarly world, and to give solid foundations for historical research, Renan undertook to publish all the Phoenician inscriptions known in his time. In 1867, he created for that purpose, with his colleagues at l'Académie des inscriptions et belles lettres, the *Corpus inscriptionum semiticarum*, of which the first fascicle, given over to Phoenician inscriptions, appeared in 1881.

As a newcomer, excavation archaeology, or special archaeology, just completed the general archaeology of surface exploration. But its objectives and methods were still spontaneous and not very carefully worked out, meeting limited needs—namely, to clear to their base the monuments partially visible in order to study them better (forms of construction, styles and decorations), to reach by shafts and tunnels the funerary chambers of the hypogeum tombs not yet violated, in order to remove the contents which would from then on be kept in museums. In the case of Renan's mission, it turned out that the monuments studied and the funerary furnishings removed were from the Persian or the Hellenistic period. It was not that Renan was interested in a particular phase of Antiquity, but as a pioneer in this field, he had to make his research on the monuments most accessible and respond to the specific expectations of the scholarly world on the intriguing question of the sarcophagi considered at that time to be Greek. It must be admitted that he had dealt fairly with the authentic Phoenician civilization, in Phoenicia itself and in one of its historical phases the best documented up to the present.

One could have thought that the discoveries of the Renan mission were going to draw the attention of the scholarly world definitively to the Persian period in Syria-Palestine and allow in this way his archaeological approach to develop: nothing of the sort happened. After a promising beginning, a long period of stagnation with no notable progress seems to have produced a disinterest in the Persian period and a certain carelessness in research. What is more, the excavations of the Renan mission had as an unfortunate effect that it multiplied the clandestine excavations and the destruction of monuments on the Lebanese coast. Until 1920, in the devious climate of treasure hunts that comes up periodically when a kind of anarchy rules, the regular excavations which could take place were themselves used to try to save the furnishings of the threatened necropolises from being plundered: this was what happened in the two funerary shafts excavated at Sidon in 1887

by O. Hamdy Bey, who sent the sarcophogi to the Istanbul museum
and published them with T. Reinach (1892).

Beginning in 1920, under the French and English mandates, archaeological work was resumed on numerous sites in Lebanon and in Syria
as well as in Palestine, but these archaeologists too were always little
interested in the Persian period. Before the Second World War, the
excavations of settlements or of more specific neighbourhoods that can
be dated to the Persian period were rare: in 1921–22, M. Pézard did
not manage to detect the least stratification, nor dwelling places of mud
bricks, on the site of ancient Qadesh (Tell Nebi Mend), in the 9 to 10
metres of deposits following the neo-Assyrian conquest; he had to be
content with some figurines and specimens of ceramic (Pézard 1931).
In 1924, M. Dunand limited himself to simple survey soundings around
the sanctuary of 'Eshmun, near Sidon (Dunand 1926). A little later, the
excavations of C.F.A. Schaeffer at Ras Shamra and Minet el-Beida,
which were essentially directed towards the vestiges of the Bronze Age,
showed only a slight interest in the Persian–Hellenistic level and nothing (or almost nothing) about it would in that case have been published,
if it had not been for a hoard of Thraco-Macedonian coins (Schaeffer
1931; 1939a: 48-50; 1939b).

Only the necropolises to some extent attracted attention, but just one
excavation was rapidly and conscientiously published, bearing in mind
the still rough-and-ready methods of the time. Under the direction of
A. Barrois, the mission of l'Ecole biblique et archéologique française
de Jérusalem undertook a search on the small tell of Neirab, south-east
of Aleppo, an Aramaean religious complex dedicated to Sahar/Sin,
whose existence was known from two neo-Babylonian stelae discovered in 1891 (Barrois and Carrière 1927). Barrois in fact excavated
the part of the necropolis left free by the modern village which occupied half the tell: this necropolis was remarkable for the length of its
utilization (at least five centuries), and the variety of types of burials
and funerary furnishings. In particular, he discovered there a small
batch of 25 cuneiform tablets, bearing on the edge some Aramaean
graffiti. Published immediately by E. Dhorme (1928) and correctly
situated in the period from Nebuchadnezzar II to Cambyses, this lot was
not, however, identified for fifty years for what it really was: a part of
the archives of a family firm closely linked to Babylon; here again, the
Persian period only appeared as a waiting place for eventual developments in research (Fales 1973: 131-42; Cagni 1990).

They were not yet interested in the little knolls as such, where there was nothing to attract the eye of the prospector apart from a scattering of broken pottery, apparently of little significance; it was almost the beginning of the 1920s before they began to grasp the significance of these humble objects which have resisted so well the test of time. Even the outcropping walls beside the sanctuaries, for example on the site of Oumm el-'Amed, still did not interest the excavators. The archaeology of occupation sites, save for specific large buildings (palaces, temples, fortresses), still remained to be created: it emerged little by little in Palestine beginning in the last decade of the nineteenth century, with reference among others to the Persian period, but it only became imperative long after.

At the time of the campaigns from 1890 to 1893 at Tell el-Hesi in the south-west of Palestine, the Englishman W.M.F. Petrie and his American assistant, F.J. Bliss, making use of large vertical sections, discovered the fundamental principle of relative (or sequential) chronology*, namely, the relation between the successive layers of archaeological deposits and the typological progressions of the pottery and other objects (Petrie 1891; Bliss 1898). Some sixty years were needed, however, in order for their intuition to lead to the stratigraphic method of dating which was going to make archaeology a scientific discipline. It meant that it would be a matter from then on of an especially demanding archaeology with upgraded methods and, as a consequence, with competent personnel and adequate equipment. The history of the excavations, especially in Palestine where they grew in number, illustrates well the difficulties that still had to be overcome in order to develop the new method and have it accepted.

Petrie and Bliss themselves did not know how to show the considerable progress to which their methodological intuition would lead, since it remained on a level that was too theoretical and did not allow them to go beyond the practice of a crudely geometrical stripping or to recognize the occupation gaps which cause a problem in reconstructing the absolute chronology*. The method of reading and graphically reassembling the real stratigraphy of the archaeological deposits still remained to be constructed. Right off, it came up against the carelessness and inaccuracy that were the rule in the hasty and skimpily supervised excavations at the beginning of the twentieth century: the most famous examples of this were provided by E. Sellin at Taanach

(1901–1904), by G. Schumacher at Megiddo (1903–1905) and especially by R.A.S. Macalister at Gezer (1902–1909) who, moreover, felt obliged to change completely the regional chronology to make a long occupation gap disappear (Sellin 1904; Schumacher 1908; Macalister 1911–1912).

One of the dangers that threatened the archaeological chronology based on this proto-stratigraphy was the temptation, so frequent in Palestine, to look for reference points for an absolute chronology based on the a priori ideas of biblicists: thus Sellin, although he knew how to learn from the unfortunate experience of Taanach by putting together a team strongly structured with architects, topographers, draughtsmen and photographers to excavate Jericho (1907–1909), fell victim to this shortcoming; he thought he had found the walls spoken of in the Bible (the wall supposedly destroyed in the 'Israelite Conquest' and the fortress constructed by Hiel of Bethel [1 Kgs 16.34]) and thus dated much too late the walls and the levels belonging in fact to Middle Bronze (Sellin and Watzinger 1913). Other archaeologists, and important ones at that, succumbered again much later, but with less excuse, to the same temptation. Nevertheless, the principal danger that threatened archaeology in this age of trial and error lay in the delay of definitive publications. Already before the First World War, the Harvard University mission to Samaria (1908–1909), under the direction of G.A. Reisner, considered a model because of its well-financed endowment, its supervisory staff, its management and its well registered and easily accessible documentation, published the results of its excavations after a considerable delay, in part attributable to the years of war. As a result, the two volumes which appeared in 1924, although remarkable for that period, did not finally play the effective role in the advance of archaeology that one had the right to expect (Reisner *et al.* 1924). Since then, the problem has only got worse: the most famous excavators, although aware of the problem, have not always been able to ensure the publication of their excavations while they were still alive. Archaeology that integrates the stratigraphic method comes up against a real difficulty: how to balance the mass of information to be handled with a reasonable delay in publication. These are two contradictory requirements needing to be reconciled, one as urgent as the other, when you realize that the excavator is always the one best placed for proposing the first level of interpretation of the objects and structures *in situ**, but that implications of every kind concerning them are tied in with the vast network of comparisons near and far which turn up from the whole regional archaeological community.

In spite of these delays, partly offset by a real competitiveness, archaeology in Palestine constructed little by little its pottery chronology through the awareness of the chain of experiences which, from one site to the other, provided its basis. But all through this phase of trial and error there remained a certain reluctance about confronting the problem of stratigraphy: as if they continued to be fascinated by the monuments, the archaeologists still preferred to let the architects have the preponderant role which they had always had, even if from then on they had to take into account the most humble structures.[4]

In this experimental laboratory which Palestine constituted before the Second World War, the archaeological excavations had hardly begun to make the Persian period emerge in the research field. In fact, the documentation of that period consisted mainly of ceramic furnishings (figurines, imported Attic pottery and a little local pottery), metallic vessels and small semi-precious objects, coming most often from tombs. The structures discovered were still few in number and their dating poorly fixed between the neo-Babylonian and Hellenistic periods. However, 21 sites would be witnesses henceforth to the Persian period, 18 of them since 1920; but only seven sites have revealed structures susceptible of interpretation, other than tombs or graves: Megiddo, Tell Abu Hawam, Tell Jemmeh, Tell el-Far'ah (southern), Lachish, Tell el-Fûl and Tell el-Kheleifeh.

Between 1948 and 1980, archaeology in Palestine made a leap forward, quantitatively and qualitatively. There was first of all a veritable explosion of research in the part that had become Israel, which was more or less controlled certainly, but which, for the Persian period alone, affected some fifty sites of all sizes. Nevertheless, up to the end of the 1960s, the excavations of the Persian levels in Palestine, despite their definite advance over the rest of the Syro-Palestine region, remained very disappointing because of the impossibility of establishing a coherent stratigraphy in the levels disrupted almost everywhere by

---

4.   We may cite as a special case the excavations of H. Ingholt at Hama in Syria (1931–38), which produced results that were relatively accurate despite the very cursory stratigraphic practices of the time. These results were certainly due to the careful records of the architect, E. Fugmann (1958), who knew how to deal with a baked-brick architecture often renovated and modified.

pits of all kinds and the work on foundations in the Hellenistic and
Roman periods.

The turning point took place when a critical study of all the publications prior to 1970 became possible, thanks to the new stratigraphic
method and the constant refinement of new excavation techniques and
a better knowledge of the local pottery. The British School of Archaeology founded by M. Wheeler had, in fact, just introduced in Palestine,
in the excavations of K. Kenyon at Jericho (1952–58) (Kenyon 1957;
1981), a new method allowing for a strict stratigraphic control of the
removal of the soil and great precision in recording finds, with appro-
priate techniques such as a grid system*, materialized in three dimensions by means of regular balks* (Wheeler 1954). Giving great importance to the vertical excavation, this archaeology undoubtedly answered
the urgent need to carry out precise stratigraphical analyses on levels
that had often been disturbed. The excavations of the 1960–70s, by
refining everywhere the diachronic reading of the remains and the
sequences of pottery, permitted the isolation from then on of the materials and the levels of the Persian period, and the beginning of the
recognition of the successive occupational phases in the depth, however limited, of these levels. From then on likewise, the old finds could
be redated and put in perspective in local typological series and in
synchronic cultural zones. Their interpetation remained, to be sure, a
matter for debate, but it was from then on possible.

Benefiting from this context, there appeared in 1973 the work of
E. Stern, in its Hebrew version, then in its English version in 1982
with a relatively limited update: *Material Culture of the Land of the
Bible in the Persian Period 538–332 BC*. The critical study and the
well thought out putting in order and by categories of the discoveries,
as they were known and clarified by the recent stratigraphic
refinement of the Persian levels, constituted the first stage in specific
archaeological research on this period, an indispensable foundation
for future developments. However, the inherent weaknesses of its time
still remained on the level of interpretation, as it relied on a foundation
of pottery evidence that was too narrow: experience today leads to a
rejection of any demonstrative value to isolated shards, always likely
to have suffered displacement between strata because of tunnels made
by animals or because of the growth of roots. On the other hand, the
absence of stratigraphies in the old publications and the frequent dis-
ruption of the levels of the Persian period stand in the way of drawing

an argument from a few shards hard to localize. Transcending the traditional separation between material and non-material elements of a culture still remains to be done, since social information can be inscribed as well in objects and architectural and spatial structures as in written sources, if we know how to find them (Schnapp 1974: 9-18); this does not mean, however, that the archaeologist can do without texts. In 1978, Stern integrated in his publication of the excavations of Tel Mevorakh a development highlighting the regional approach, an approach already practised, but one which was soon going to come into general use in archaeological research. The following year he emphasized the interest in this approach, which opened up for archaeologists new perspectives on the social, economic and political structures of societies considered in their own area (Stern 1978: 79-80; 1979: 34-39). Tel Mevorakh in the Persian period as well as in the Hellenistic period was just a secondary agricultural settlement in the area of Sharon, dependent upon Dor. The dependent connection of this site with Dor appears evident if we make its abandonment correspond, as the archaeological results suggest, to the displacement of the administrative centre from Dor to Caesarea at the time of the Herodian foundation, which was probably accompanied by a regional restructuring.

As archaeology, since the time of Petrie, became more and more attentive to any movable object or fragment of object and to its multiple meaning, then to traces of human presence and activities, it was led to establish teams of competent technical staff (anthropologists, zoologists, pottery experts, numismatists, chemist-restorers, metal specialists), whose competence increasingly flows from scientific disciplines. With the method of systematic stratigraphy developed by the British school, archaeology has risen to the rank of a science. It has dealings today on an equal footing with the life sciences and the natural sciences, at the same time on the level of laboratories (controlled analyses phase) and in the course of excavations (phase of extraction and pre-diagnostic appraisal). Besides the anthropologist and zoologist (ethno- and zoo-archaeologist) who are already consulted on the archaeological sites, sedimentation geologists, pedologists, and botanists who specialize in the study of particular finds (pollens and larger remnants such as grains and coals) become full-fledged members of the excavation teams. The scientific study of all the material from the ancient sites, initiated by the prehistorians, reaches today the whole archaeological world, making archaeology more and more dependent on the

exact sciences. The extraordinary scientific development of archaeology in its methods of analysis and its problematics cannot, however,
make up for the development of solid research programs (Schnapp
1974).

In the case of Transeuphratene in the Persian period, these new research perspectives are not purely theoretical or distant, since excavation work of this kind is already under way. We will cite in particular
the Israelite–American excavations at Tel Michal, the Phoenician coastal
site in southern Sharon, near Jaffa.[5] The archaeological programme of
Tel Michal follows quite as much from the new multidisciplinary
broadening of research as from the geo-archaeological problematic
which will be presented in the following chapter. The publication of the
four campaigns of excavations carried out from 1977 to 1980 on this
site already gives a quite remarkable portrait of its history, from which
there stands out in particular a spatial distribution of functions which
runs across the centuries, even over periods of abandonment.

Now, the Persian period is documented in an exceptional way on this
site spread over five hills: up to six levels of occupation have been per-
ceived on the hill called 'central tell' or 'high hill'. It was there that a
fort seems to have been periodically reconstructed from Late Bronze I
(about 1500) until the Hellenistic Period; the three eastern hills would
become cult places beginning with the Iron Age. The same occupation
outline can also be applied, it seems, to the Persian period. At the turn
of the sixth and fifth centuries, the site seems to have experienced an
occupation of a semi-permanent type: granary pits (for cereals?) and
an area of temporary dwelling places, perhaps with tents, since domestic ovens were found without associated permanent structures, while a
large structure in unfired bricks could fill the function of protection
on the site of ancient forts. Then, in the middle of the fifth century,
the three cultic sites would have been reorganized, while the enormous
grain silos would have been constructed on another hill, next to a vast
agricultural structure which could have served as a warehouse. On the
southern slope of the lower city, a potter's workshop, containing five
storage jars crushed on the ground, promises an interesting study of
local pottery and the techniques in clay, given the place accorded in
the programme to the geological study of the environment. Finally, a
necropolis from the end of the Persian period is made up of at least

5.   Herzog *et al.* 1980; Herzog 1981; see also Herzog *et al.* (eds.) 1989, which
we have not yet been able to consult.

120 tombs already excavated. The reports of the preliminary excavations can help us get some idea of the variety of studies that we have the right to expect from such a programme, such as the relations between public and private buildings, farmers and craftsmen, localization and nature of harbour installations, the significance of the appearance of the necropolis at the end of the period, and so on.

The broadening of the field applying to archaeological research and of the scientific collaboration to which it leads is again illustrated, for the period which interests us, in various fields of which we will cite at least one example, namely, underwater and coastal archaeology, a fast-developing area. It inevitably raises complex problems linked to the modifications in the shorelines and, as a consequence, is led to integrate into its research the geomorphology of coastal regions. H. Frost had, since 1973, given a significant example of the integration of the two disciplines in regard to the ports on the Phoenician coast; despite the stopping of research on these sites, other work illustrates since then a fruitful interdisciplinarity, already well established.[6]

If the sites of intense occupation and specific activity remain privileged locations, rich in significant remains, the space, any space where human activity took place across the centuries, it too humanized and structured, is an important witness, open to analysis by geo-archaeological methods. Thus, today another form of archaeology, called 'spatial', is essential, in the search everywhere for strategies adapted to regional clusters which it deals with as a whole. Since the 1970s, it is the complementarity of these two types of research, excavations and surveys, which is opening up new perspectives for the future. Programmes integrating them are presently under way in Syria-Palestine and some of them, such as that at Tel Michal of which we have just spoken, involve the Persian period.

6. Frost 1973; Dalongeville and Sanlaville 1980; Sanlaville 1978; Collombier 1987; 1988; Raban 1983; Raban (ed.)1988.

Chapter 5

THE COMPLEMENTARITY OF GEO-ARCHAEOLOGY

Geo-archaeology, or spatial archaeology, forms part of historical geography, although it is a recent extension of the archaeology of sites responding to a new historical questioning. Historical geography has old qualifications to exploit, since it already constituted one of the main lines of the investigation of Herodotus in the middle of the fifth century before our era. Ethnographic as well as historical, it was also the basis for the great historical work of Diodorus of Sicily, in the first century BCE. Likewise, the identification of precise regions and places was the dominant geographical preoccupation of the interpeters of the Bible, Jewish and Christian pilgrims come from distant communities into these Near Eastern regions won over to the Greco-Roman way of life, in which Greek onomastics tended to conceal the old Semitic place names. In the same way, the *Onomasticon* of Eusebius of Caesarea (fourth century CE) served as the basis for the research of Jerome (fifth century), who hurried to translate it into Latin: 'Just as those who have seen Athens understand Greek history better', he wrote, '. . . in the same way the Sacred Scriptures will be taken up with greater insight by those who have seen the land of Judah with their own eyes and who will have known how to recognize the cities and the ancient places with their names, just as easily those which have changed (names) as the principal ones (which have not changed)'.[1]

We have already touched on the enthusiastic resumption of learned journeys to the Near East in the first half of the nineteenth century, and of the search for illustrious places of Graeco-Roman and biblical antiquity even before the beginnings of the modern epigraphic corpuses. That form of knowledge, at the same time geographical and historical, stamped at the end of the nineteenth century by Positivism*, became at

---

1. Jerome, *Praefatio in libros Paralipomenon* 423A; Maraval 1988: 348-49 (other writings of Jerome on this theme).

the beginning of the twentieth century a scientific discipline in the full sense of the term, divided up according to linguistic fields and geographical regions, thanks to the epigraphical corpuses and the first excavations then under way. At that time they gave it the name 'historical topography', which always surprises the geographers a little, but is easily explained when we realize that the places mentioned in the military and commercial itineraries of the ancient texts are often located in relation to one another in terms of orientation, of spatial or temporal distance (according to the way of travel, which was sometimes stated) and of natural obstacles, all notions that could be transferred into the content of topographical maps. 'Topographical history' is, in fact, the correlation–identification of place names attested in the ancient texts with the places (sites, districts and regions) defined by modern topography and designated by name in current or recent toponymy. This discipline has its own methods of analysis and correlation, whose principles of application are recalled by J. and L. Robert in the particularly difficult case of Anatolia (Robert and Robert 1977: 50-63).

For Palestine, highly favoured in this research, the geographical and historical knowledge at the end of the nineteenth century greatly exceeded the topographical inventory of the identified sites, thanks to the Arabic or Arabized place names which had so often preserved the structure of ancient names; already beginning in 1838, E. Robinson and E. Smith had become famous for their inventory of these names (Robinson and Smith 1856a; 1856b). In 1865, the Palestine Exploration Fund was formed, which had as its objective to carry out research in the 'Holy Land', in archaeology, geography, geology and natural sciences. It began by having a topographical and descriptive survey drawn up, precise enough to attempt to resolve the problems of territorial boundaries, such as those of the biblical book of Joshua, with regard to the tribes of Israel (North 1979: 124 [with bibliography]). This first far-reaching project was realized for the whole of Cisjordan by C. Conder and H.H. Kitchener, resulting in a monumental publication: seven volumes of text and 26 maps at 1/63,360 in 1880–84 (North 1979: 125 n. 13 [with bibliography]). All the problems of the famous territorial boundaries were, for all that, certainly not resolved, but this systematic work served as a base for later cartographic surveys, in particular for the German military surveys during the First World War, then for the production of a map at 1/100,000 by the administration of the British Mandate. That map has become a classic because of its relative accuracy, its handy scale and coordinate grid ('the Palestine grid'), which still serves today as an international reference for all the works on historical geography in Palestine. This brilliant beginning by Conder and Kitchener was immediately followed up, with remarkable results, since three historical topographic maps of Palestine appeared, one right after the other, in 1890, 1892 and 1894.[2] The last one was much more than an atlas map provided with a historical commentary: due to the Scot, George Adam Smith, it accompanied a historical geography whose value was rapidly recognized worldwide.[3] Its author considered himself the heir of almost a century of systematic research in 'Syriology', having to do just as much with numerous aspects of geographical science, including observations on social life, as with the classical and recent disciplines, such as epigraphy and archaeology. Beyond the 'historical topography' that could not be ignored, which he used in a critical manner, with imagination and common sense, Smith had attempted (and succeeded very well for that time) to restore to history a global geographical dimension. The constant value and modernity of this dimension of physical and human geography, presented as a horizon for archaeological research, is what we wish to emphasize.

Yet, at the dawn of the archaeological era, up to the 1950s, the historical geography of the Near East hardly tempted geography specialists. For Palestine, it had remained the work of biblical scholars, become amateur geographers in order to render an account of their from now on intimate knowledge of the country. To get an idea of the profound familiarity that the biblical and archaeological Institutes established in Jerusalem since 1890[4] maintained with the populations settled in the country and on its desert fringes, and with their way of life, it is enough to skim through the numerous annual reports of their study journeys, the reports of explorations and of specific missions which weaved as they went along a network always closer and larger

2.   The first map was made by H. Fischer for the biblical scholar H. Guthe 1911, who revised it: see North 1979: 127.
3.   Smith 1966 (the work went through 25 editions from 1894 to 1931 and was again re-edited in 1966).
4.   The first was the French Biblical and Archaeological School, founded with the name of Practical School of Biblical Studies by Père M. J. Lagrange in 1890. Next came the American School of Oriental Research in 1900, the German Evangelical Institute in 1901, then the British School of Archaeology in 1922.

at the same time. Before the days of the automobile, which needs a network of highways, study itineraries done on horseback were especially flexible and effective: 'You see less, but you see better', the biblical scholar F.M. Abel remarked. Despite the numerous works of geologists, climatologists, hydrologists, pedologists, botanists, zoologists, and the administrative investigations on demography and economics, the geographical syntheses on Palestine carried out by geography specialists remained rare before 1955, and were of little significance, and especially made no notable contribution to historical geography. It was to Abel that we owe the only authoritative work in this field during the first third of the twentieth century: the *Géographie de la Palestine* in two volumes appeared in 1933 and 1938. He integrated in the best way, for the time, an exacting historical geography and the most up-to-date knowledge on the various physical and bioclimatic milieus concerned. As well as its great historical value, its cartographic precision and toponymic richness, equally remarkable, the author applied to it in a systematic manner the new principle of historical topography, theoretically exact, but difficult to apply, which the famous American archaeologist W.F. Albright practised and recommended: no biblical site can be situated with any probability in a place that did not supply the pottery corresponding to the biblical period in question.

At the end of this period there appeared in 1957 *The Geography of the Bible* of D. Baly, an Anglo-American geographer with a good biblical background. This book just confirmed the intuition of Smith on the physical and human peculiarities of each of the sub-regions of Palestine. The geographic constants (environment, natural resources, economy and traditional way of life) were described there in a striking manner and abundantly illustrated with biblical texts, but the historical topography leaves something to be desired. Five years later appeared a new historical geography which just renewed and brought up to date Abel's historical topography and made up for the weakness in Baly's book: *The Land of the Bible, a Historical Geography* by Y. Aharoni, an Israeli archaeologist. Published first in Hebrew, this book marked an epoch from its first English edition in 1966. Furthermore, thanks to his work in the basin of Nahal Beersheba, Aharoni introduced into archaeology the regional approach, from then on practised everywhere; in this way he could set out on an adequate scale the problems of a socio-ethnic nature which preoccupied him. Thus there followed, one after another, up to a recent date, syntheses of historical geography

on Palestine, centred sometimes on geography, sometimes on history
and archaeology, but all preserving the main theme given by the biblical texts themselves concerning the regions or 'cantons' of Palestine.

With regard to the other regions of Syria-Palestine, no work of synthesis has attained the level of historical knowledge of the *Topographie
historique de la Syrie antique et médiévale* of R. Dussaud, which appeared in 1927. Its principal objective was to identify the sites of ancient
place names: following, in his case too, region by region, the thread
of ancient and medieval descriptions of itineraries and administrative
districts, relatively well defined, given the physical differentiations of
the country and its compartmentalization, especially in mountainous
regions, he successfully managed some striking diachronic crosscheck-
ings. Nevertheless, his geographical descriptions remained too brief
or only allusive. The geographers J. Weulersse and R. Thoumin, who
shortly after published their studies of human geography in different
cities and regions of Syria, hardly managed to catch up to the problematics of historical topography as presented by Dussaud, there,
where one could expect them to do so, on the terrain itself.

In his book published in 1971—*Syrien, eine geographische Landeskunde*—E. Wirth was the first geographer to integrate carefully
historical and archaeological research in this region. A long chapter
on 'the heritage of the past' allowed him to underscore space-time
relations and human resources and dynamics.[5] In his descriptions of
the different sectors of Syria, more vast than those of Palestine, the
author was able to integrate the results of some archaeological work,
principally ground surveys and aerial surveys. Now, the geographical-
historical relation which this manual illustrates does not function for
the benefit of geography alone, but archaeological research in the broad
sense finds in it a basic source, an indispensable geographical analysis
for understanding the ancient potentialities, both physical and human,
of the regions studied.

This journey seems to us illuminating. Straight off, two strong points
stand out, which permit the situating of archaeology in relation to
historical geography. We establish in the first place that there has
always been a historical geography, sometimes centred on the identification of places with a secure mooring place in a past which has faded

---

5.   This chapter is comprised of two especially important subchapters: 'The
Historical Roots of the Present Agrarian Landscapes' and 'The Population of Syria in
its Historical Differentiations in the Spatial and Social Order'.

in the memory of local populations, sometimes reached by way of 'humanized' milieus which preserved the trace of successive legacies in terms of structuring (occupation and utilization-management) of space. In the second place, the scale of regions and sub-regions is undoubtedly imperative in Syria-Palestine as the most productive approach in historical geography. That regional approach, affirmed more or less in research for almost two centuries, is henceforth essential for archaeology. In short, archaeology should also be geographical. It was definitely that, through instinct, back when it was in its infancy in the age of discoveries of the exotic Orient; it must be it consciously today, by fully assuming a complex problematics with all its implications on the plane of methodology and means. Beyond the excavated sites that reveal the most dense structures of human activities in the past, there are the ancient environments, marked by activities linked to the preceding, which still remain to be discovered.

A third piece of evidence leads us to the same conclusions. The identification of places corresponding to ancient place names remains a theoretical game, with little reliable historical results, if the toponymic or 'topographical' procedure, already complicated in the field of linguistic developments and transfers, is not controlled by a minimal geographic knowledge of the ancient structures of occupation and use of space, still possible to locate in the region under consideration. These successive occupations, attributable to various historical phases, must still correspond to the socio-political and economic systems which had produced them, at least insofar as we can reconstitute them using the ancient texts. In many cases, the 'historical topography' even then only offers at present stepping stones for historical research.

As an example of a thorny problem in this field, we can cite the historiographical texts of the Bible, all marked with ideological or theological a priori principles, some even plainly legendary, such as those which deal with the origins of ancient Israel, at least up to the consolidation of the monarchy and the administrative and scribal structures indispensable for control of the facts by writing. Biblical scholars, aware of the necessity of 'proofs' from archaeology to identify the places mentioned in these texts, have a tendency to leave it up to archaeologists. Now, the latter, aware of their inability to become proficient in literary criticism, are relatively little concerned with problems of identification; in fact, they can only formulate on this subject risky and ultimately dangerous hypotheses as they see these transformed in secondary sources into so-called 'commonly accepted learned opinions' (Miller 1983: 125-28). As a result, neither the biblical scholars, nor the archaeologists can take on separately the responsibility for the necessary research in this field, since the identification of many biblical place names necessitates interdisciplinary studies and amounts to a long, drawn out process. For certain place names, attested only in contexts that are not very clear, the undertaking is furthermore even premature or straight out impossible.

The biblical books of Ezra and Nehemiah, relating to the Persian period in Palestine, principally in the 'province' of Judah, can serve as an example. We know that they are relatively close to the historical and geographical realities that they touch on, but they do nonetheless introduce an ideological vision which compels the researcher to practise a real literary 'stratigraphy' to come closest to the realities. Thus there must be isolated, in a double text (Ezra 2 and Neh. 7.6-72), what applies to 'the inhabitants of the province', listed according to their places of residence, in the midst of lists of names of families/clans and individuals having no other connections than Jerusalem. The geographical distribution of these clearly localized inhabitants recalls that of the ancient tribe of Benjamin, limited in the north and south, from the area of Bethel-Ai to that of Bethlehem-Netophah, but it differed from the tribe of Benjamin in its east–west axis. It extended, in fact, from the oasis of Jericho to the last hills and the small valleys opening on to the coastal plain that characterize the area of Lod, Hadid and Ono; this area with a special and still ambiguous administrative status could have extended the ancient tribal area towards the west. The distribution of the 'inhabitants of the province' (of Judaea), identifiable thanks to the villages mentioned in this text, definitely reminds us of the ancient territory of Benjamin, but enlarged in the direction of the east–west communication routes that predominate in this area: the problem is to know to what this distribution corresponds.

Another spatial structure indicated in the book of Nehemiah (3.9-19), which seems to apply to groups or partial groups whose status is not made clear, employed here in statute labour under the direction of their respective rulers, for the reconstruction of the walls of Jerusalem, presents a north–south axis corresponding to that of the crest-line. This personnel exploitable for large projects could have belonged to state administered property, like the group mentioned two verses earlier (Neh. 3.7), if, however, it is really a matter in this difficult passage of people at the service of a governor of Transeuphratene and working in the area which was reserved to him.

These biblical place names a priori represent spatial structures quite well brought out in the texts and well situated in the regional geography despite the uncertainties that remain with regard to the identification of some of them. In this case, the logical development of the research should proceed, in our view, by way of a detailed investigation of a geo-archaeological type, which would attempt, of course, to identify the sites corresponding to the place names still difficult to localize, but would especially endeavour to reconstitute the structures of the occupation and use of the milieus involved, in order to verify whether they correspond to those suggested by the texts. Such an investigation would, at the very least, permit the placing of the still unresolved problems of historical topography on better foundations, and especially the enlarging of the problematics of a regional historical geography, thanks to the new methods and stragegies of geo-archaeology. It will be enough to mention some of the means and methods of research used in the Earth sciences (remote detection*, geophysical* or geochemical* prospecting), as well as some research strategies used in the different branches of geomorphology, most of which are particularly suited for the study of complex spatial systems.[6]

The still preliminary phase in which archaeological research in Transjordan happens to be provides every opportunity for the geo-archaeological approach to define the problems to be confronted. In fact, the image of turning in on themselves presented in the Persian period by all four regions, Edom, Moab, Ammon and Galaad, reminds us of a double climatic and human fact which characterized them periodically—namely, on the one hand, the reduction to a narrow band of the possibilities for pluvial agriculture and even for pastoral resources, springs and pasturage, in periods of prolonged drought, and on the other hand, the development of the socio-political system of integrating nomadism* analysed by M.B. Rowton (1980 [with bibliography, 292 n. 2]), all the more operative here, since this band is cut up into isolated units because of the nature of the topography. This double fact, of a geographical nature, seems to us to be true for Transjordan as a whole, but in a differentiated manner according to the conditions of the physical milieu in each of the regions.

---

6.   See an example of the adaptation of these methods and strategies to the geo-archaeology presently practised in France in Beeching 1990 and Brochier 1990.

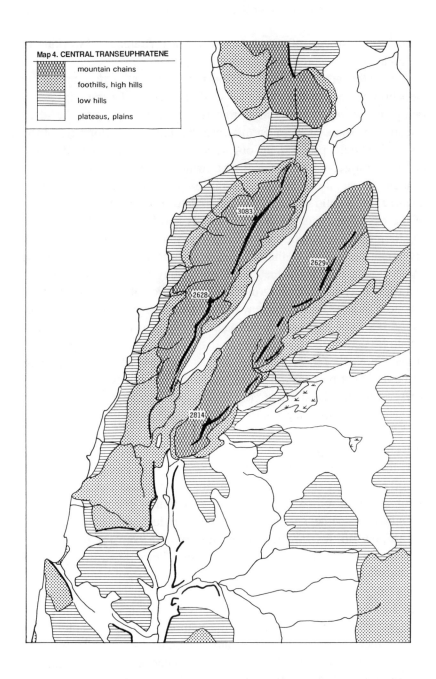

Map 4. CENTRAL TRANSEUPHRATENE

- mountain chains
- foothills, high hills
- low hills
- plateaus, plains

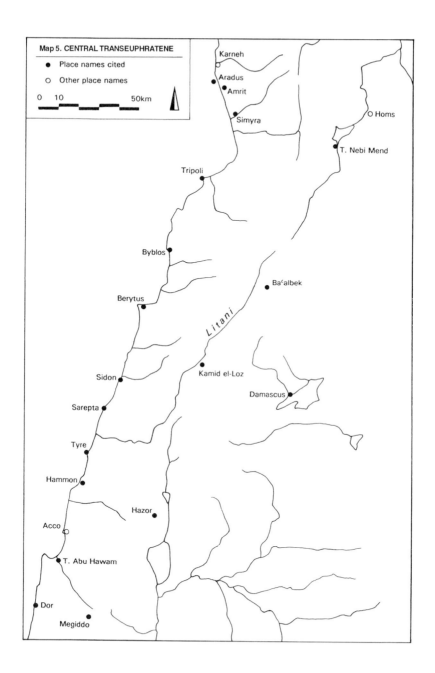

Map 5. CENTRAL TRANSEUPHRATENE

● Place names cited
○ Other place names

0   10   50km

Karneh
Aradus
Amrit
Simyra
Homs
T. Nebi Mend
Tripoli
Byblos
Baʿalbek
Berytus
Litani
Sidon
Kamid el-Loz
Damascus
Sarepta
Tyre
Hammon
Hazor
Acco
T. Abu Hawam
Dor
Megiddo

In the region of Edom, most open to the desert and therefore the
least favoured, sedentary life seems to have disappeared in the Persian
period on the southern plateau, but persisted in the north, in the hill
country (Hart 1986). This physical and human setting, which needs to
be refined considerably more, makes it possible to think of the presence of integrating nomadism which can have, just as easily as a local
government or the Imperial administration, the integrative political
role that we see in the centralized form taken by the mining activity in
the area of Feinan in the Assyro-Babylonian and Persian periods
(Hauptmann *et al.* 1986; Hauptmann and Weisberger 1987: 428).

For the rest of agro-pastoral Transjordan, to the north of Edom, we
must begin by differentiating the Moabite plateau, largely open towards
the steppe, from the limestone and more or less wooded massifs that
follow it to the north. Moab seems to disappear from the archaeological and historical map in the Persian period,[7] whereas Ammon and
Galaad present a completely contrasting picture. Ammon seems to
have preserved its sedentary population and probably its prior agricultural and military structures, if not a political and administrative
organization even, inherited from a vassalized kingdom in the setting
of the Assyrian and neo-Babylonian Empires. Organized for its own
self-protection in depth, since Iron Age IIB at least, the period of the
vassal kingdom, the Ammanitide apparently resisted the nomadic pres-
sures very well, thanks to its local defence or the Persian army, and
maintained its external relations all during the Persian period.[8]

What was it like at that time in the mountain-refuge of Galaad? The
upsurge which had marked the period from Iron II (Braemer 1992)
up to and including the neo-Babylonian period, seems to have been
followed by a downturn and perhaps by a real hiatus in sedentary
occupation: was it a simple withdrawal into settlement structures that
are not very evident, or was it pressure from nomadism following an
agricultural overexploitation of the mountainous and wooded surround-
ings, or just ignorance of the local pottery, our only criterion for dating

7. At the northern boundary of Moab, the sites of Heshbon and Iktanu are to be
connected to the Ammanitide.
8. Knauf (1990), attributes the round towers in the region to the Persian defence
system confronting the nomads. See Khalil 1986 and Hadidi 1987 for the beautiful
perfume burner in bronze and the imported Greek pottery coming from a tomb to the
West of Amman, evidence of the external relations of the Ammanitide in the Persian
period.

and very difficult to use in the absence of imports or other external criteria? We cannot decide among these factors in arriving at an interpretation, but in the end they could be cumulative. In any case, the
second hypothesis could gain weight if it should be proved that there
was a real climatic deterioration from the beginning of the Iron Age.
In fact, the mountains of Galaad had presented ever since the third millennium a competely deforested eastern slope, 'steppified', open to nomads as far as Jerash/Galaza; the whole mountain area must have been
sufficiently deforested since Iron II to have become, in the Persian
period, deeply penetrable, in the absence of a defensive system maintained or favoured by a unifying government, to tell the truth something difficult to conceive. The only system of defence known up to
the present for that period was found in the Jordan Valley, precisely
opposite the outlets of the two valleys of the massif of Ajlun, at Tell
el-Mazar and at Tell es-Sa'idiyeh. The first site, which seems to have
been a military and administrative centre, is situated facing the valley
of Wadi Rajib, but could have also controlled movement in the main
route of Nahr Zerqa, the Jabbok of the Bible (Yassine 1983). The second site, nearer the Jordan, seems to have kept watch over a ford and
the valley of Wadi Kufrinjeh whose basin runs up to the midpoint of
the massif as far as Ajlûn, where the principal circulation routes cross
the mountain.[9]

As these examples show, a historical geography integrating human
structures and ways of life appears to be the most efficacious approach
for formulating a problematic for archaeological research adapted to
the preliminary level at which we find ourselves in Transjordan. This
problematic must be diachronic and regional at this stage of the research, that is to say, it should on the one hand integrate the stratigraphic excavations developed more or less horizontally with skilled
and controlled surveying and on the other hand include a good part of
the first millennium BCE. It would in this way be meant to differentiate the Persian period from the surrounding periods, as much on the
level of movable objects such as ordinary pottery (poorly identified
and necessitating regional studies of comparable stratigraphy) as on
that of the structuring of space (distribution and function of the sites

---

9.   Pritchard 1985. The nature of the occupation in the Persian period at Tell
Deir-'Alla, facing the outlet of Nahr Zerqa, is still uncertain; it will, in any case, have
to be related to that of Tulul edh-Dhahab, a double tell situated at a strategic point of
that transverse valley (see Gordon 1987).

in keeping with the agricultural, pastoral, mineral or other resources, and with the regional economic flows and networks).

Although the works of Dussaud and Wirth set up the historical geography of Syria-Lebanon as a whole on good foundations, the Persian period remains, in this discipline, a poor relative when it comes to the greatest part of this geographical area, namely, the interior of Syria, which is almost unknown in comparison with the coastal regions. But it is especially the imbalance in knowledge and research programmes between the two zones that seems to us to be prejudicial to the understanding of the relations that existed between them, above all in a historical perspective. The regional approach of some archaeological programmes in inner Syria, including the Persian period, lets us envisage the future with confidence however, since it will contribute, as it has already begun to do, to a restabilizing of knowledge.

We will especially draw attention to a problematic sometimes mentioned by historians, which has never been elaborated in archaeological programmes, still less in carrying them out, and which is worth considering. For a long time a contrast has been set up, using many legitimate arguments, between the coastal area of Phoenician cities and the non-urban and essentially arameophone areas of inner Syria. But archaeology and linguistic geography confirm the existence between these two cultural areas of relations, for which we understand neither the networks nor their basic socio-economic systems: rather than a clear-cut frontier, our meagre documentation indicates instead a complex interpenetration in the areas of intramountainous communication. We can gauge all the interest that would be produced by a precise geoarchaeological programme from the Mediterranean coast up to a corresponding well documented sector of inner Syria. Research of this kind has begun to be carried out on a part of the Homs Gap by way of a preliminary survey, carried out with the least expense (Sapin 1989b). Its provisional results already situate the deforestation of the mountains, its agricultural development in stages beginning with a pastoral economy, its dense village structure, preserved without great change up to the present, its internal networks and the commercial and military network which runs through it, from the Assyro-Babylonian period to the Hellenistic period, in a homogeneous Aramaic toponymic context, in which, however, the important role of the Phoenician city of Arwad is recognizable. Inevitably, this survey leads to extensions beyond that area, at least as far as the valley of the Orontes.

The written sources relative to the Persian period make Transeuphratene, and particularly Syria, seem like a territory for military manoeuvres and occasionally for confrontations; there we read of revolts of local governments and satrapies, repressions, reconquest of Egypt, Persian interventions in Greece and Cyprus. The armies of the governments of the moment unceasingly cut across this region and used it without fail as a logistical base: however, we know nothing, or almost nothing, of this function (Galling 1937: 44-47; Rainey 1969). Since inner Syria was made up of villages and was non-urban, would it not be administered, at least in part, from the royal and military spheres, in which the Persian authorities would have had as an objective to further the progressive settling of a part of the native populations? What socio-economic and political organization characterized these populations and other peoples eventually resettled there? It would be necessary to confront resolutely the spatial dimensions that problems of this type imply and, as the research necessarily goes through archaeological programmes capable of providing new documentation, it would be advisable to include, around the centres that a priori exercised politico-administrative functions, the spaces that they administered. It matters little that the Persian period has left in certain areas only humble rural villages more or less sedentary or nomadic, much less spectacular than the urban and monumental ruins. These meagre traces, by their spatial and temporal distribution and density, are themselves indicative of historical organizations and human decisions. Geoarchaeology, which borrows its concepts from the new geography, takes care of all the traces left by human activities which can be spotted, and carries out an analysis of them in order to discern the ancient systems of organization that helped them fulfil their specific functions.

Since it is impossible to study all the sites and the whole areas around them with the same care, it is evident that some choices are necessary, based on priorities. Among well-tried spatial approaches, that of transections or strip-transections* work very well for regional groups, as well as on smaller scales, especially when a minimal documentation of maps and aerial photographs in different scales is available. It allows in actual fact for the transition from the regional scale to the local scale* and vice versa, which is indispensable in any geographical research on natural or human processes, involving extremely varied spatial and temporal scales. If the archaeological sites require specialists from the social and natural sciences, the regional programmes of geo-archaeology especially need, among other specialists, geographers for the integration of physical data with human data and the interpretation of the process of dynamic interaction. From the choice of the analyses having priority all the way along until the synthesis, geography is much more than an auxiliary for archaeology: it is the necessary partner at every stage in the whole programme of archaeological research including geo-archaeology.

This discipline presents one final advantage, still more sensitive in a region where the concerns of the present put the historical patrimonies and the archives of the soil in danger.[10] It is to this that all the national Antiquities services, more and more consciously, have recourse; it lets them end up with the best results and with the least cost to construct emergency programmes rapidly and make an inventory of the threatened reserves, taking on in a way the equivalent of a salvage excavation, on a regional scale and no longer on a merely local scale. Geo-archaeology could perhaps even make it possible to go further and envisage the idea of archaeological reserves as they have already set up nature reserves for flora and fauna.

---

10. This danger is all the more real since there does not really exist any 'archaeological vacuum', especially in countries with ancient civilizations: see Renimel 1979.

Chapter 6

OF WHAT USE ARE INSCRIPTIONS?

The underdevelopment of studies on the ancient Near East in comparison with other fields of research on Antiquity is explained in part
by the fact that they are always essentially dominated by two disciplines that F. Braudel would have described as 'imperialist', namely,
archaeology, of which there has just been question, and epigraphy.[1]
One or the other of these two disciplines, and preferably both, constitute an obligatory avenue to becoming a researcher recognized by one's
peers in the field of the ancient Near East. However, in actual fact this
sectarian point of view is not completely unfounded—for example, in
the case of epigraphy, since the inscription represents a direct and
sometimes explicit witness to the epoch under consideration and, by this
fact, a fundamental element in the historical investigation.

Now, the close study of Northwest Semitic inscriptions involves far
more difficult procedures than those for Greek or Roman inscriptions:
since Greek and Latin are relatively well-known languages, most of
the inscriptions are in translations accepted by almost all the specialists
and any divergences in interpretation often only relate to minor points;
at a pinch, one could feel relatively secure in using these translations,
without verifying them for oneself. Nevertheless, this approach is often
considered risky. But in the field of the ancient Near East, it is far more
risky to make do with translations of Northwest Semitic inscriptions
since they are susceptible of very divergent interpretations, due to the
fact that they have neither vowels, nor, most often, separation marks
between words (Elayi 1990a): ultimately, two translations of a same
inscription can diverge to such an extent that reading them does not
even leave room for identifying them as the same text. What is more,
the corpus of Northwest Semitic inscriptions is quite restricted and the

1. Epigraphy brings into play paleography (study of writing), philology (study
of forms) and linguistics (study of language).

known vocabulary limited: for example, in 1983 W. Röllig managed with difficulty to compile a list of some 668 words for the whole Phoenician–Punic field (Röllig 1983: 376); the new inscriptions discovered since then have not increased this number in a spectacular way. One understands, then, the usefulness of closely studying epigraphy oneself in such a difficult documentary context.

Among the written documents relative to Transeuphratene in the Persian period, there is one of them so exceptional that in the opinion of the majority of researchers, it eclipses all the others: it is very definitely the Old Testament, of which we will speak at length in the following chapter. Despite the multiplication of epigraphical discoveries, many biblical scholars still have a tendency to consider the inscriptions, especially if they are not monumental, as documents of quite secondary interest. There are numerous studies on biblical exegesis today that do not even mention the inscriptions contemporaneous with the texts that they are analysing. Nevertheless, the unending exegetical squabbles that we continue to witness clearly show that the results of literary criticism are not always certain and reveal the weakness of the historical testimony of the text in comparison with inscriptions.

In actual fact, there are inscriptions and inscriptions. Traditional epigraphy is especially interested in monumental inscriptions, that is to say, inscriptions engraved, most often carefully, on a stone medium, which generally have a certain length and concern important personages occupying a special rank in the society of their time. These are, for example, official, royal or civic inscriptions, or else votive or funerary inscriptions whose authors are not necessarily politicians, but quite well-to-do people who could pay for this type of engraving. The advantage of these inscriptions is to inform us principally about 'official history' and about the powerful and rich of a given society: thus, the funerary inscription of King 'Eshmun'azor of Sidon teaches us about the nature of his government, the political and religious functions of the queen mother, and the expansion of the city's territory by the gift of imperial lands (*KAI* 14; Elayi 1989b: 38-41, 52-53). But this type of inscription ought to be subjected to an especially rigourous textual criticism, since they are often written to glorify an important person, while living or after his death, or to spread political propaganda. The resources of epigraphy alone rarely suffice to establish an exact translation of the inscription, since a judicious choice about the meaning of words cannot be made through a simple recourse to philology or linguistics. Rare, however, are those who, like O. Carena, begin to deplore the considerable backwardness of Near Eastern historiography in comparison with that of Greco-Roman sources. He writes, for example:

> the works about ancient Near East history appear even today as syntheses in the style of the nineteenth century. They lack an historicistic reflection *qua talis*, and tacitly accept the presupposition that the ancient Near Eastern annalist gives us documents which can already be taken into consideration as such and brought together into a wider perspective or amalgamated with data from other sources and fields, especially archaeological. This state of things continues not to be perceived as being inadequate.[2]

Despite the pronounced preference that some epigraphers still have for the monumental inscriptions, we acknowlege all the same the great merit of Northwest Semitic epigraphers in being interested too, for about the last two decades, in non-monumental inscriptions which they had too long wrongfully neglected. In fact, necessity is not extraneous to this healthy orientation: the number of monumental inscriptions being greatly limited, much more so than that of Akkadian inscriptions, for example, the epigraphers had to search elsewhere to find material for their research and have resorted to collecting as well more or less brief inscriptions whose medium was less 'noble' than stone. These non-monumental inscriptions can be very varied in nature: ostraca*, monetary inscriptions, countermarks* and graffiti, inscriptions and graffiti on bowls or other objects, stamps and various markers.

The ostraca constitute a separate category, which contrasts in every respect with the monumental inscriptions, but whose historical significance is not necessarily less:[3] they are to a slight extent the ancient equivalent of our scratch pads. They offer the immense advantage of providing direct testimonies since they were always contemporaneous with the events that they recorded and are often dateable, by stratigraphical methods if they have been discovered *in situ*, by paleographic study, or by the circumstantial details that they sometimes contain. They are in principle not very suspect of trying to disguise the truth, since they had no literary pretensions nor any ideological objective; they were not destined for posterity since the period of their use was limited; they were never put on public display, but were reserved for

2. Carena 1989: 5; for Assyriology, see Garelli 1964: 31-40, 120-26.
3. Lemaire 1977: 13-15 (the ostraca assembled in this book are prior to the Persian period).

a limited and private usage. They reflect some aspects of the daily life of the period in which they were written and can touch on extremely varied subjects: school exercises, lists of pilgrims or debtors, for example. Thus, ostraca bearing lists of proper names have been discovered close to the wall of the monumental podium of the temple of 'Eshmun in Sidon (Vanel 1967; Elayi 1989b: 46-48, 64-65): even if speculation about the purpose of these lists goes on all the time, onomastic study makes it possible to know what kinds of visitors frequented this temple from the fifth to the fourth century BCE. An ostracon with a Phoenician inscription of seven lines was discovered in the excavations of Acco in 1980; it was an official order for vases of different categories, placed with a group of artisans, for the person in charge of the temple (Dothan 1985).

In regard to monetary inscriptions, their study fairly recently provides a new subject for research (Elayi 1991a; see also Chapter 9). Their principal advantage is that they are most often dated, either explicitly or else indirectly, offering in this way valuable evidence about the issuing authority at the time and the representation that it wanted to give of itself to the local literate users. Thus, on the coinage of 'Aynel of Byblos, dated to the middle of the fourth century, the substitution of the name of the city in place of the royal title—'Aynel, king of Byblos'—was the sign of a major political change, characterized by the weakening of royal power to the advantage of the civic community (Elayi 1987b: 43). There still remains much to be done for this category of inscriptions, especially on the level of the interpretation of Phoenician abbreviations and of the reading of the inscriptions on the so-called 'Philisto-Arabian'* coins (Elayi and Elayi 1988; Mildenberg 1992), to which we will return in Chapter 9.

The Northwest Semitic seals, which are most often stamped seals (scarabs, scaraboids and conoids), have been known since the last century through the works of M. de Voguë, M.A. Levy, Ch. Clermont-Ganneau and M. Lidzbarski (Lemaire 1988 [with bibliography]); in the twentieth century, especially beginning with the 1950s, the multiplication of discoveries and publications of unpublished seals has led to the issuing of catalogues of inscribed ones. But it is especially in the last fifteen years or so that this type of seal has become the object of an attentive research, no doubt stimulated by the increase in their numbers and the growing interest that they arouse in the antiquities market. The number of seals listed is on the whole quite limited, however, since it must today amount to about a thousand examples, of which only a fraction date from the Persian period (Avigad 1970; Bordreuil 1986). The study of these seals is difficult, since they are dispersed in numerous public and private collections, and the place of their discovery often remains uncertain. In general they bear the owner's name, patronymic and office; they could have been used for sealing contracts or other private acts, but they could also have had an official use in administration: an example of this would be the fiscal seals probably used in the collection of the tithe in the various localities of the territory of Tyre (Greenfield 1985). Some seals bear, besides the inscription, a figurative representation, especially revealing the personality of the engraver, since it was often engraved before the sale, the name of the owner only being inscribed at the time of purchase. Sometimes, instead of the seals themselves being found, only the imprints of seals, or bullae*, on clay for example, come to light (Avigad 1976). We must mention too the stamps on amphorae which generally bear the name of the maker, invaluable information for identifying the workshop where it was made and for retracing the trade route followed by the receptacle up to the place where it was discovered. To finish up, we must note the imbalance that exists between the numerous studies on Hebrew seals, bullae and stamps, carried out in a biblical perspective (for example, Avigad 1988), and those which concern other seals, whether Phoenician, Aramaic or Ammonite, still quite poorly known.

In addition to their official inscription, coins can also be countermarked, that is to say, marked with a punch to add a simplified iconographic motif or a very brief inscription: the official or private nature of the usage of these countermarks is always very much debated, but the matter is hardly likely to make headway as long as a systematic study of countermarks is not carried out.

Finally, inscriptions painted on vases and graffiti incised on vases, coins or other objects constitute a research field that is still little explored, but one that is promising: they can provide various pieces of information on the content, the capacity or utilization of the vase, or, which is most frequently the case, on the owner of the object. Thus, the site of Shiqmona (Sykaminon), at the foot of the Mount Carmel spur, has yielded several jars or fragments of jars with inscriptions painted in black ink: the most complete ones mention the owner (son of Matton), the date in the local year (in the twenty-fifth year of the king), the content (fermented wine), the quality of the wine (good) and its place of origin (Gat Carmel) (Delavault and Lemaire 1979).

At present we can assess at what point it is useful to take into account all the inscriptions together, whose importance in that case is not based on the category to which they belong, with the result that they would provide information that is often complementary. The fact that Northwest Semitic epigraphy is not very plentiful offers the undoubted advantage of also being able to take into account the briefest inscriptions, whereas the Assyriologist, for example, has far too many inscriptions to study to be able to consider them all;[4] but such a situation presents on the other hand a grave inconvenience: it too frequently leads the Northwest Semitic epigrapher to act like a 'treasure hunter'. His only goal would seem to be the discovery of an unedited document, an activity in other respects very difficult because of the often hazardous conditions for prospecting on the ground, and especially because of the scarcity of the inscriptions compared to the number of the specialists searching for them. The least inscription with a few letters becomes the subject of an article exaggerating its importance, and an inscription of more than a line becomes a veritable scientific event, worthy of a presentation at the Académie des inscriptions et belles-lettres.

In fact, the publication of an inscription is usually limited to a paleographical and philological study and a brief commentary called 'historical'. With this publication finished, the researchers start off without delay on the quest for another unpublished inscription; in a situation of a scarcity of unpublished inscriptions, which is not rare in our field, they ceaselessly republish the same inscriptions. It is true that this field of study presents numerous problems and that these successive republications often set in motion boundless energy and ingenuity to try to improve the editions of documents, but is there not something better to do for the 'treasure-hunting' epigraphers, temporarily empty-handed? It would be more useful, for example, as the specialists of Greek epigraphy quite systematically do, to group together the already published inscriptions into corpuses by regions, by periods or by themes, provided with good indexes for consultation. Such corpuses would offer the immense advantage of a rapid and easy handling of

4.   Thus, 30,000 tablets of the *Annals* of Assurbanipal still await translation in the storerooms of the Britsh Museum in London.

the data, with various inscriptions put in series, which would be indispensable for their historical interpetation. It is especially deplorable that the excellent *Corpus des inscriptionum Semiticarum* (*CIS*) undertaken by the Académie des inscriptions et belles-lettres in the time of Renan, was not completed and scarcely interests today's epigraphers any longer; everything goes along as if Northwest Semitic epigraphy had become split up, individualized, without any planning of the whole thing nor any dynamic collective enterprise. Sometimes catalogues are published that group together a particular collection, or are presented as selected passages, but the corpuses, even limited ones, are very rare: we may cite, however, the small corpus of the Phoenician inscriptions of Palestine, which contains 62 such inscriptions, and that of the Northwest Semitic coin graffiti, which groups together 51 graffiti (Lemaire and Elayi 1987); and we will mention two corpuses, only in part epigraphical, which have been announced: that of Northwest Semitic seals and of Phoenician coins of the Persian period.[5]

To tell the truth, epigraphy is a difficult discipline to practise because it is very 'specialized' and requires a permanent investment on the part of the researcher. It is a discipline as well which has acquired today great precision and strictness; but at the same time, it has remained too fossilized and isolated from the other social sciences, not without consequent repercussions. Thus, it is not enough to study in minute detail the form of the letters of an inscription and to date it, in order to deduce from it the type of writing used in such a place in a given period. It is also necessary to isolate the local trends in the writing, by distinguishing technical constraints of every kind likely to modify the tracing of the letters, constraints eventually imposed on the scribe by the sponsor of the inscription, the writing habits and the personal choices of each scribe, and the accidental aspects of some written forms.

The most conscientious epigraphers are content with some cursory remarks on these problems, whereas there is needed an in-depth study of all the technical problems that come up on the level of the process of writing, the tools and the material used, the purpose of the inscription and the identity and motivations of the sponsor. For example, the frequently irregular form of the round Phoenician letter *'ayin* in the engraved inscriptions should not be interpreted as a local tendency in

---

5. Corpus of seals (Avigad 1970; 1976; 1988; Bordreuil 1986); corpus of coins (Elayi 1993). We should note too the interest in epigraphical bibliographies: Suder 1984.

the writing: it suffices to know how an engraver has trouble in producing a circle with a graver, to understand that the irregularity is due exclusively here to the technical difficulty (Elayi 1991a). In the same way, in some monetary captions from Byblos of the fourth century, we notice that the upstrokes of the letters which project above the line shorten themselves or bend themselves to the right: once again, it is not a local development in the writing, but a solution discovered by the engraver to resolve a problem of a lack of space.

It is risky, too, for the epigrapher to venture a historical interpretation of an inscription, or even at times to propose a translation, using just the epigrapher's view of things, without resorting to other social sciences. Let us take the example of an inscribed seal recently discovered at Sarepta; if we confine ourselves to strictly paleographical and philological study, the translation that automatically comes to mind is the following: '(property) of the king of Sarepta'. An irreproachable translation, certainly, from the syntactical point of view, but a historical countersense, given what we know of the socio-political structures and the territorial organization of the region in that period; the context makes it necessary then to reject this natural translation and to look for another, even if it is not evident; for example, '(Property) of the king. Sarepta' has been proposed (Elayi 1989b: 89,101).

Not only is epigraphy liable to make scientific errors by isolating itself from the other social sciences, but its principal error is to wish to set itself up as an end in itself. We have no intention of criticizing research on unpublished documents, on condition however that it goes beyond the simplistic phase of a 'treasure hunt'. Unpublished documents, and especially written ones, are indispensable in making our knowledge of Transeuphratene progress, but how many published documents remain unexploited? In other words, what is the purpose of lining up the pieces of a puzzle if nobody tries or is concerned to piece them together? When they publish previously unpublished documents, some epigraphers remain convinced that they are the only ones advancing studies in their field, confusing in this way the end and the means; they forget that a document is not an end in itself, but simply a tool which, if used judiciously with the intersecting information provided by other disciplines, can permit our knowledge of the past to make progress. Other epigraphers ease their consciences: the studies are still too embryonic in our sector to permit a historical approach; let us pile up the documents to fill in the gaps and, later, 'do history'! Through thoughtlessness, laziness or incompetence, they refuse to open up the closed world of epigraphy.

Certainly, the information that an inscription can provide on the form and evolution of writing and on those of language are extremely valuable, but we have the right to expect other things from them on the socio-cultural context of writing. It is essential to wonder about the profession of scribe and the different specialties of the scribes depending on the medium of the inscription and its destination, namely, for public or private use. Thus, the engraver of seals for private use is distinguished theoretically from the engraver of dies* in a minting workshop, who was a sort of employee of the state, and on the other hand, the technique of the engraver on metal differs from that of the engraver on a semiprecious stone. In reality, if we consider, for example, the Phoenician minting workshops, they were probably devised in as informal a way as the contemporary Greek workshops, in the sense that the artisans could also, in their own workshop, carry out the official orders for engraving monetary coins. Some engravers of seals or goldsmiths could sometimes engrave coins, particularly when the issuing authority had emergency coinage struck, and when, as a consequence, monetary production had at the same time to be increased and speeded up.

It is interesting too to wonder whether the scribe knew how to read: if pushed, he could copy the model without understanding it; in the case of engravers of coins who had to write backwards on the coins, the frequent forgetting of the inversion of the letters would be due to the custom of writing in the right direction and proves then that the author of this error was proficient in writing. What was the place of the scribe in society? Why did the engravers of Phoenician coins, for example, never sign their works as many of their Greek counterparts did? Was it because this post was considered a public service and the personality of the artisan took second place?

The study of writing should make it possible to measure the degree of literacy of the society in which it was practised. For that, it must be asked who ordered the inscription, for what readers and to transmit what message. In the case of the Phoenician coastal cities, a certain number of clues allow us to think that a quite large part of the population was literate. In fact, if the various authorities issuing Phoenician coinage had taken the trouble to have an inscription on their money, when the decorative symbol would suffice for identification, it was

because a large part of the users were able to read. It was the period
when monetary graffiti began to be developed (Lemaire and Elayi
1987), at the same time as graffiti on vases: a graffito was the work of
anyone who took a hard point and wrote his name, or anything else,
on an object. The development of cursive writing in the fourth century probably reflected the extension of writing with ink on papyrus:
nothing remains of this perishable material in the humid soil of the
Levantine coast, but the practice of writing in ink is confirmed on a
few ostraca which have come down to us (Millard 1991). As a working hypothesis to confirm this, we can presuppose some development
of literacy in the Persian period (especially in the fourth century) in the
urbanized coastal centres, whereas in some interior regions of Transeuphratene, tradition of an oral character probably still very much prevailed over that of writing.

The study of writing can also teach us about the process of cultural
integration: thus, a Graeco-Phoenician seal of the Hermes the Shepherd
type bears, it seems, the divine name 'Baal', the first two letters of
which are written in Phoenician and the third in Greek.[6] We can wonder about the significance of this mistake in writing, without any doubt
involuntary on the part of the engraver, but it shows clearly that the
engraver was acquainted with Greek writing. We can also study the
links between writing and magic, for example from the mysterious
signs that are mingled at times with the Phoenician monetary inscriptions. We are very pleased that the most recent Symposium organized
by the University of Liège in Belgium on Phoenician writing concerned itself not only with its formal aspects, but also with the problems of its diffusion; with the relation between the oral and the written;
and with the connections between writing and society, writing and
culture, and writing and magic;[7] it is to be hoped that the research on
Transeuphratene in the Persian period will be increasingly represented in the new methodological approaches of this type.

The historical contribution of the content of an inscription can at
times be considerable, but the so-called 'historical' commentary that
secondarily completes the study of the inscription is in general incapable of making this contribution come to light, because the monolithic

6.   Elayi 1988: 152-53 n. 81; we cannot see why the inscription would not be
contemporaneous with the seal.
7.   Baurain *et al.* (eds.) 1991.

approach of a discipline which has too great a tendency to be 'imperialist' rarely sets out a historical problematic correctly and knows still less about how to propose solutions. For example, a study of monetary inscriptions from the Persian period, far from limiting itself to a paleographical and philological commentary, should call on competent social scientists to consider, particularly, the numismatic, iconographic, political and economic aspects (Elayi 1991a). Properly designed, this study would be bound to help draw up a geo-political map of local governments and provide information on the forms of government, the relation between small states and the great powers of the time, and the nature of local economies and trade.

To finish up, let us mention that the Northwest Semitic inscriptions constitute the majority, but not the sum total of the epigraphical documents on Transeuphratene in the Persian period. The royal Achaemenid inscriptions also provide some succinct pieces of information, for example, the charter with a trilingual structure (Old Persian, Elamite and Akkadian) from the palace of Darius I at Susa.[8] A certain number of Greek or Cypriot inscriptions, for the most part bilingual (Greek and Phoenician), contain masses of useful information. The most important is the Athenian decree in honour of the king of Sidon, Straton (probably 'Abd'ashtart I), engraved on a stele of marble from Pentelicus which was discovered on the Acropolis, beside the Parthenon (*KAI* 60; see Elayi 1987b: 52, 113; Baslez and Briquel 1989); the text is made up of two parts: a decree of Cephisodote, an Athenian orator and politician, concerning Straton and his descendants, and an amendment proposed by Menexene, applying to the Sidonian negotiators staying in Athens. This inscription enlightens us particularly about the intermediary role played by Sidon in the relations between Athens and the Persians, on the forms of expression of the proxeny* that Straton made use of at Sidon for the Athenian foreigners, and on the status of the Sidonian negotiators staying at Athens. What we have just said about the Northwest Semitic inscriptions essentially applies to these non-Semitic inscriptions, apart from the differences that, in these other sectors of research, epigraphy is generally not considered an 'imperialist' science and that there exist some corpuses in which these inscriptions

8. See above, Introduction, n. 2.

are classified in a practical way. But they must be used with much more prudence and discernment than the local inscriptions because they come from cultures exterior to Transeuphratene, with different mentalities, which must be taken into account.

Chapter 7

READING THE TEXTUAL SOURCES ANOTHER WAY

Before asking ourselves how to use the textual sources, let us remind ourselves that we are dependent on the documentary corpus, as it *de facto* exists at the present time. The textual sources on Transeuphratene in the Persian period are much more abundant than the epigraphical sources and split up into two main categories: the classical texts and the Old Testament.

The classical texts provide lots of diverse data, direct and indirect, whose nature depends on the focus of their authors' interest and on their available information. Some subjects are satisfactorily developed, while others, just as important from our modern point of view, are not even skimmed over: thus, there is very much information in regard to most of the populations of the Levantine coast, with whom extensive international trade made it possible for the Greeks to have close contacts; we will cite the example of Herodas of Syracuse who, according to Xenophon, was at Tyre for commercial purposes when the Persians undertook huge naval preparations: right away he took the first ship leaving for Greece and went to give the Lacedemonians a detailed report on what he had seen (Xenophon, *Hellenica* 3.4.1).

But the texts remain practically silent on the populations of the hinterland, even of that close by: probably the Greeks had hardly ever met on the sea routes or in the ports of the Aegean or Levant these populations which did not have any hankering for the sea. The Greeks would, however, know some of them, who were engaged in extensive interior trade, in caravans or on rivers, probably because they had been able to have direct contact with them in the large Levantine ports: it was in this way that Herodotus provides some information on the caravan trade of the Arabs which he would have been able to meet for example at Gaza, where his wanderings had taken him (Herodotus 3.5-7). Before the Hellenistic period, we have no evidence to show that the

Greeks were acquainted with the Hebrews and their political and religious characteristics (Will and Orrieux 1986: 67; Momigliano 1987: 13-31). It was because of this that Herodotus has no clear idea of the populations making up Palestine and never mentions the province of Judaea; in the same way, the Greeks are mentioned just a few times in the Old Testament, under the form *Yawan* (Ionians) or *Kittim* (from the name Kition in Cyprus) (Herodotus 2.104; 7.89; Gen. 10.4).

The Greeks were, on the other hand, deeply interested in all aspects of the Persian Achaemenid Empire: its political history, its institutions, its administrative organization, central and provincial, and its customs, as is indicated in particular by the numerous *Persika* writings of the fifth and sixth centuries, of which only some fragments have come down to us, such as those of Deinon; in the same way, the Hellenistic authors make numerous references to Persian customs. One of the main centres of interest for the Greeks was, without any doubt, the political events involving Greeks and Persians, in particular the Median Wars, which provided the basic points of Herodotus's work, and of numerous developments in Thucydides, Isocrates, Diodorus, Plutarch and Pausanias, for example. Greek historiography was also interested in the institutions and customs of the Persian Empire in a perspective that was above all comparative, in order to show off to advantage the corresponding Greek institutions and customs (Briant 1987): we may cite, for example, Herodotus, Thucydides or Xenophon. Other authors such as Plato or Arrian used comparisons of an ethnographic nature to make concrete to their readers such or such a physical feature of the territories of the Persian Empire. The classical sources can provide geographical information as well, as is the case for two sources of a completely different nature: the *Periplus* of pseudo-Skylax of Caryanda, about which there is agreement that it should be considered a reconnaissance of the coasts of the Persian Empire carried out for Philip II of Macedonia, and the *Geography* of Strabo. It would seem, on the whole, that the genuine curiosity of the Greeks about the Persian Empire in the fifth century had given way, in the fourth, to a stereotypical view, fed by a more accurate knowledge of details (Kuhrt and Sancisi-Weerdenburg 1987: xiii).

For too long a time, modern historians have given the classical sources an indiscriminating confidence and a credulity for which the Hellenocentric vision of Antiquity has been, most often, responsible. Thus, the distorted vision of the Greek victory at Salamis, presented by Greek authors such as Herodotus as the turning point in Greek history, making complete the triumph of civilization over barbarism and of democracy over despotism, has been widely repeated up to our own time, particularly under this simplistic form: if Salamis had been a Persian victory, modern democracy would perhaps not have existed (Young 1980, especially 239). While we can understand the biased point of view of the Greek authors about these events, who, like Herodotus, did endeavour at times to be objective, it is difficult to accept the lack of objectivity of modern historians. The Persian perspective on the Median Wars was much different from the Greek perspective: the conflict with the Greek cities was, as a matter of fact, quite marginal for the Persian sovereigns, more preoccupied with the surveillance of Babylon which had just revolted in 482 BCE, or with their endemic conflict, with Egypt. On the other hand, the Persians did not have a monopoly on despotism, which many a Greek city had experienced, beginning with Athens; they even favoured some democratic governments in the Greek cities of Asia Minor; in a word, the Persian Empire had nothing of a barbaric Empire about it, as people have too long been led to believe.

The injudicious use of the Classical sources can also be explained by ignorance or the rejection of critical studies on classical historiography, in particular concerning the structure of the account and its literary form. Critical study can detect the progressive distortion of the account of the death of the king of Sidon, Strato ('Abd'ashtart I), through the successive narratives of Theopompus, Elien, Anaximene of Lampsaque, Maxime of Tyre and Jerome (Elayi 1989: 23-24, 181). The testimony of Theopompus, a contemporary of the event, quoted by Athenaeus and repeated by Elien, joins that of Anaximene of Lampsaque, a contemporary of Alexander and perhaps also of the event. Even if the contemporary character of a witness is not a guarantee of authenticity, it still offers the possibility of direct information, and the convergence of two contemporary testimonies merits our attention. On the other hand, the moralizing accounts of Maxime of Tyre and of Jerome really distort the event: according to the first author, Strato would have died destitute and Nicocles of the effects of his imprisonment; the misogyny of Jerome made him link the death of the king of Sidon to the intervention of a woman, in the context of the failure of the second revolt of the satraps, with which this event had nothing to do. We may say finally that the critical examination of the

classical sources cannot be done without a good knowledge of the corresponding Near Eastern sources, which is not self-evident to a specialist in Classical studies, even one disposed to taking all things into
consideration.

If the uncritical use of the Classical sources is not very scientific,
the opposite approach, amounting to a systematic denigration, is just
as unacceptable. That attitude, very fortunately more rare, can be
explained, without excusing it, as a reaction to the preceding attitude.
It comes from specialists on the Persian Empire, or Orientalists with
somewhat narrow views. As A. Kuhrt and H. Sancisi-Weerdenburg
wrote recently (Kuhrt and Sancisi-Weerdenburg 1987: ix), not without a sense of humour, the Median Wars still continue today, bringing
face to face Hellenocentrists and Iranocentrists! We could note analogous situations in other fields of Orientalism, for example in Phoenician studies, where the fact of having been a Hellenist before having
become an Orientalist constitutes for some an indelible defect, whereas
it should represent a supplementary guarantee of objectivity and
openmindedness. In actual fact, the local sources should no longer be
used with uncritical confidence, but would gain from being submitted
to a decoding in the same way as the Classical sources. It is true,
however, that these local sources have priority since they have one
less deformity: they do not reflect the point of view of external cultures. Can we accept the jest of T.C. Young, according to whom we
would be better off not to have sources at all than to have poor sources
(Young 1980: 237)? Let us say right off that sources, even poor ones,
alway constitute a source of information, even if only about the mentality of their authors. On the other hand, there exist many subjects
for which the Classical sources constitute our only avenues for information; thus, the role of the Phoenician fleet in the Persian naval forces
is known only through these sources (Elayi 1989: 161-95): it is up to
the historian to know how to use everything, even the so-called 'poor
sources'.

It is certain that the use of Classical textual sources involves risks
and their limits are undeniable, but their importance as documentation
on Transeuphratene in the Persian period is evident, on condition of
knowing how to read them differently, that is to say, to make each
piece of evidence undergo a critical examination and to submit it, in
particular, to a rigorous structural and formal analysis and to a meticulous ideological deciphering (Briant 1982: 491-506), doing one's best

not to introduce a modern ideological apriorism. It would be advisable first of all to identify the literary genre to which the source belongs and take into account all the rules of this genre: thus, a literary narrative can use real historical personages, but make them develop in an imaginary setting, or, conversely, make imaginary personages evolve in a real historical setting (Griffiths 1987). The personality and mentality of the author, his motivations and objective must be studied in the text, too: despite Herodotus's efforts to be objective, his account of the Median Wars is to a great extent reorientated by his admiration for the Athenians and by his apologetic presentation on the role of the Delphic oracle (Elayi 1978–79). We would have to determine as much as possible the source (or sources) of each piece of historical information to be able to measure its degree of authenticity. The classical authors made great use of oral testimony, to which they seemed to attach the utmost importance; even if some of them, like Herodotus, maintained that they did not believe everything told them, they apparently did not engage in the internal criticism carried out by today's historian: it is desirable in this case to have at one's disposal other documents to control the collected testimonies.

Access to archival documents, even explicit, should, as far as possible, be controlled as well. Access to the public chronicles of Tyre seems to have been permitted, according to Flavius Josephus, who quotes Menander (Josephus, *Apion* 1.107.111; *Ant.* 8.55). At the time of his voyage to this city, carried out to gather information for his inquiry on Heracles, Herodotus was, however, content, for a reason which we do not know, to limit himself to carefully chosen informants—the priests of Melqart/Heracles—which raises, incidentally, the problem of the language of communication:

> I began a conversation with the priests of the god; I asked them how much time had passed since the establishment of their sanctuary . . . they replied that this sanctuary had been established at the same time that Tyre was founded, and that Tyre had been inhabited for two thousand and three hundred years (Herodotus 2.44).

On the other hand, a regular visitor to the Achaemenid court like Ctesias (Diodorus 2.32) seems to have had access without difficulty to the central Persian archives for his historical information. The role of oral transmission remained of primordial importance, even in the consultation of archives, which could be done through the intermediary of local informers. Thus, when Herodotus went to inquire in Egypt, the priests read him a book: 'After [Min], the priests cited, according to a book, the names of three hundred and thirty other kings' (Herodotus 2.100).

The critical analysis of the classical sources can benefit at times from an unexpected documentary situation. Let us take the case of the information according to which the Phoenician city of Tripolis was formed by three fortified cities, which is found in many authors such as pseudo-Skylax, Diodorus, Strabo, Pomponius Mela and Pliny the Elder.[1] No local source makes it possible to verify this and it has generally been concluded that it was an etiological explanation, invented as an afterthought, to explain the name 'Tripolis', which means in Greek, 'three cities'. But it may be shown, thanks to comparisons with some archaeological discoveries, with ancient accounts of travellers and with the testimony of Arab historians from the Middle Ages, Al-Baladury and Ibn Al-Atir, that the Classical sources were exact: Tripolis was formed in the Persian Period from three towns, each surrounded by a fortification.

Finally, it has often been said, especially by those who show interest only in new documents, that the Classical sources have been appealed to so much that they have nothing more to say. In fact, if the sources no longer say anything, it is because they are always asked the same questions. As P. Briant has aptly stated, 'It is sometimes not so much the sources themselves as the ability of the historians to ask questions that is threatened by exhaustion' (Briant 1982: 491). The multi- and interdisciplinary approaches that we are attempting to develop and encourage in order to take a new look at Transeuphratene are bound to ask new questions of the Classical sources, which are better used, if possible, in the original text, since the translations can always be blamed for misinterpretation.

The Old Testament represents the second category of textual sources on Achaemenid Transeuphratene. It comes to us under its canonical* form as a series of three collections of books which are not the books of authors, despite the appearance of some of them: the Pentateuch* (or *Torah*, 'Law'), the Prophets and the Writings.[2] It is, in fact, a matter

1.   Pseudo-Skylax, *Periplus* 1.104; Diodorus 16.41-42, 45; Strabo 16.2.15; Pomponius Mela, *Chronography* 1.67; Pliny the Elder, *Natural History* 5.17, etc.; see on this subject, Elayi 1990a.

2.   It was already this way at the beginning of the second century BCE, according to the prologue of the book of Jesus Ben Sirach (a book of an author this time!)

of a collective work, due to groups of scholars whom we can define through their various social functions (storytellers, priests, royal and other scribes, prophets, etc.) and their traditions of theological schools, throughout the history of ancient Israelite society. This literary work is a work of long duration, which was progressively formed during almost a millennium. It was the result of an organic development of basic texts, belonging to the most varied genres: songs, stories and sagas of clans or of tribes, cult legends, collections of legal judgments, sacrificial rituals, sacred instructions, prayers, hymns and lamentations, royal annals, historical, prophetic and didactic accounts, wisdom instructions, visions, oracles, and so on.

These texts, which represented in the beginning something that emanated directly from Israelite social life through centuries, have come down to us woven into compilations, reworked many times and always made more complex, right up to forming books and collections of books. There are, for example, juridical collections of sentences and laws, codified and integrated into the historical narrative of the Pentateuch; the words of the prophets, transmitted by their disciples, annotated and reconstructed into the so-called prophetic books; or, again, individual or collective psalms, reused in the liturgy of the Jerusalem Temple or other cultic places, developed and collected together in the book of Psalms.

Although the original elements, in the course of the redactional process, were progressively detached from the particular situations and events which had created them, historiographical concern was not missing from them. In a society sensitive to narrative genres, it would be normal for its literate people to produce historical works, even though they would be of mixed genres, such as the Pentateuch, the Deuteronomistic history[3] and the work of the Chronicler;[4] to this latter have been attached the two books of Ezra and Nehemiah, which seem to have

However, the Jewish tradition of the Middle Ages subdivided the prophetic books into 'Former Prophets', called today 'historical books', and 'Latter Prophets', which consist of the large collections: Isaiah, Jeremiah and Ezekiel and the twelve minor prophets.

3.   For an introduction to the Old Testament, see that of Rendtorff 1986: 278-316.

4.   The two books of Chronicles, centred on the monarchical period and on the Jerusalem Temple, certainly constitute a self-contained history, although a late touching up was aimed at having them followed by the double book of Ezra and Nehemiah, a work clearly distinct from the preceding because of its centres of interest and its less priestly theological vision.

taken their definitive form towards the end of the Persian period or shortly after (Williamson 1987: 45-46). These latter two books, which partly intersect, present two segments of the period which concerns us: the first segment from 538–515 BCE according to the point of view of the book of Ezra (520–515 BCE according to most modern historians), deals with the return from the exile and the reconstruction of the Jerusalem Temple, to which the prophetic books of Haggai and Zechariah also bear witness; and the second segment, beginning in 458 or 445 BCE, describes the mission confided by the king of Persia to Ezra and to Nehemiah, whose chronological order is still very much debated. The redaction of the books of Ezra and Nehemiah is relatively close to the period of which they speak, but for all that they cannot be taken as objective or neutral reports. Like all the biblical books, these two books are bearers of theological messages: therefore, the exegete concerned with historical criticism will have to rediscover the socio-political and religious points of view which inspired them.

In fact, almost all the canonical books[5] have undergone an intense redactional activity in the course of the two centuries and a half (586–333 BCE) called by biblical scholars 'the exilic and postexilic periods', and provide information on the new state of affairs of the descendants of the ancient Israelites, whether they were dispersed in the Empire or gathered together at Jerusalem and in the small province of Judaea. If the scriptural phenomenon was so active during this period, it was definitely because it fulfilled an essential function for the dismembered community, which was largely uprooted, submissive to a foreign power and deprived of its own political institutions—namely, a function that provided an ethno-cultural and religious identity and a fundamental structuring for that community. The return from exile of some deportees, the reconstruction of the Temple and the walls of Jerusalem, the socio-religious reorganization of the small province of Judaea through the missions of Ezra and Nehemiah—all these historical facts which took place against the backdrop of the Achaemenid government constituted new conditions for existence whose stakes should be balanced with those of the Diaspora. In view of this common structuring, the

---

5. Only the books of Qohelet (Ecclesiastes) and Daniel were composed in the Hellenistic period. However, it is not impossible that redactional activity affecting many books of the Old Testament, including the Pentateuch group, had been carried out up to the beginning of the Hellenistic period (about 300); see Crüsemann 1989.

new redactional activity had to devote itself to responding to the fundamental needs of both the returnees and the Diaspora, at least on the religious level. The traces of these at times contradictory concerns are met with in all the books of the Old Testament.

We recognize right off that these traces are often difficult to identify and interpret, by reason of their non-explicit form and because of their theological character. Only a better knowledge of the populations living in Judaea and in the Diaspora, as well as of the other populations subjugated by the Empire, can throw light on our historical reading of the biblical texts, rather than the reverse. It is important to keep this in mind and to use the Old Testament with great prudence, as a body of distinct and indirect accounts. Guiding threads exist, however: if the canonical books as a whole are strongly marked by the exilic and postexilic periods, it is the Pentateuch which constitutes the touchstone of the whole critical approach to the Old Testament. For more than a century, in particular since J. Wellhausen (Wellhausen 1883), Old Testament exegesis has brought to the fore the fact that the starting point for all research on the history of ancient Israel is situated in the context and in the texts of the postexilic period, which were so preoccupied with problems of cult. From his examination of the cult that was at the same time anthropological and historical (place, ritual, feasts, personnel and resources of the priestly class), and despite his negative judgment about the postexilic period, considered a period of religious decline, Wellhausen was led to the history of the concepts of the *Torah* and theocracy which, in his eyes, made it possible to differentiate nascent Judaism from ancient Israel and which are essentially set out in the Pentateuch.

The fruitfulness of this regressive historical approach has certainly been the reason for the lasting favour enjoyed by the critical method of exegesis which went hand in hand with it, namely, the 'documentary hypothesis' or 'theory of sources' (Wellhausen 1899). The system of Wellhausen was completed by other literary critical approaches such as that of H. Gunkel who, through the identification of literary genres, became interested in the traditions and in their roots (*Sitz im Leben*) (Gunkel 1910; see also Gilbert 1979): the common position of these two authors was their interest in the most ancient texts and in the preliterary traditions. Conversely, the new criticism, always dealing with the analysis of the Pentateuch, concentrates today on the process

of composition and redaction, which explains the new historical interest
in the exilic and postexilic periods.

Today more than ever we realize that the scriptural fact itself, so
complex in its formation and so fundamental in its implications, certainly constitutes the essential connection between the Old Testament
and the Persian period. Already well developed as early as the end of
the monarchical period at Jerusalem and more than ever at work in
the milieu of the Babylonian exile,[6] this reality from then on assumed
a founding value and a normative character. The Pentateuch is the
example-type of this, in the strong sense of that phrase. The promise
to the ancestors, the distinctive signs of the religious covenant, the
mythical history on the beginnings of the world, of humanity and the
people of Israel, including the divinely led and protected wandering
from Egypt to Canaan, and the successive juridical codes (Code of the
Covenant, Deuteronomy, Law of Holiness), all these elements which
founded the Judaic community and let it maintain its religious and
community bonds despite the dispersion and the absence of political
independence, became, during the two centuries of the Persian period,
Scripture and Law of a new kind.

This Scripture was indispensable as a norm for the dispersed Judaic
community, by the very will of the Great King who granted the imperial authorization, 'the recognition of local norms by the authorities of
the Empire', according to the formula of P. Frei (Frei and Koch
1984: 13). It is difficult to ascertain whether the *Torah* promulgated
by Ezra at Jerusalem, with jurisdiction over all the Jews of Transeuphratene (Ezra 7.25), was already the whole Pentateuch, or only a
part of it, or just its preliminary draft. Anyway it had the title and
authority of a royal law (Ezra 7.26; see Crüsemann 1989: 345-50).

---

6.   Modern historical criticism (see de Pury and Römer 1989: 48-66) has brought
down to the period of the exile two of the traditional sources of the Pentateuch (J and
E) which Wellhausen (1899), and then Noth (1967 [1943] and 1960 [1948]),
seemed to have definitively set at the beginning of the monarchical period. However,
there is agreement on the fact that important elements of the historical and prophetical
books, some Psalms and some Proverbs, had already existed in successive editions
all through the second part of the monarchical period, therefore before the exile. On
the other hand, the Babylonian exile had surely been the most fruitful moment for
recapitulating theological works: the Deuteronomistic history (from Joshua to Kings),
the combined JE work (Genesis to Numbers), the large-scale collection of prophetic
oracles, etc.

Furthermore, the habit which the Persian kings had of treating as irrevocable all their written and sealed decrees seems to have led scribes like Ezra to practise an additive, recapitulative writing, without subtracting any of the ancient juridical writings. Finally, the very content of the Pentateuch shows its underlying agreement with the little that we know of the social and religious situation of the Jews in Judaea and in the Diaspora, who found themselves, for example, facing the necessity of respecting Persian power and of setting aside therefore every vestige of messianic prophetism.

The books stemming from the prophetic movement did not have this normative status and could express more freely the real expectations of the community and its opinions, together with subversive echoes, as well as the internal social and religious tensions. Thus, the three last books of the collection of the twelve minor prophets, Haggai, Zechariah and Malachi, oriented the community towards the expectation of a decisive divine intervention—namely, a messianic king, a preparatory coming of a final prophet, a new Jerusalem, apocalypse (in the literal sense), an expectation exacting social morality and cultic service. We find the prophetic imprint of this postexilic period of Judaic structuring in most of the other prophetic books, particularly in the three principal ones: Isaiah, Jeremiah and Ezekiel. The book of Isaiah, for example, is made up of three parts, of which the last, which is postexilic, clearly agrees with the prophetic messages that we have just mentioned; but in addition to this supplement to the first two parts (from the eighth century and the end of the exilic period), the redactional additions show that these last anonymous Isaian authors combined the theological themes of judgment and salvation in order to appeal to the postexilic community, so that it might make right and justice prevail.

The group of 'Writings', in which are found the books of Ezra and Nehemiah and Chronicles, gathers together works of different periods, including the Hellenistic period, and the most varied genres. From the Psalms to the historical novel of Esther and Job and Proverbs, the numerous historical roots of these books can be found apparently in the Persian period. Apart from the short didactic narrative of Ruth and the Song of Songs, which seem to belong to the wisdom literature and which give a distinguished role to the woman,[7] probably indicating

---

7.   The role of a woman is also the central theme of the book of Judith, which is not part of the Jewish canon, but has been retained in the ancient versions (Septuagint,

a protest against the establishment,[8] most of the books are clearly composite from a redactional point of view and necessitate a literary analysis (linguistic, structural, stylistic) and a historical critical analysis as complex and closely argued as for the prophetic books and the Pentateuch.

All in all, the great majority of the texts give evidence, at least on the level of their final redaction, of theological and socio-religious positions which asserted themselves in the context of the 'Judaic' community that had been shattered, but was on the way to being restructured in the Persian period. The historical books which deal more or less directly and clearly with the Persian period (Ezra and Nehemiah and some passages of Haggai and Zechariah) have no distinguishing literary form. Only a literary-critical examination, often very complex, applied to all the texts, can make it possible for the basic elements, the clusters and the redactional links to stand out. A historical-critical examination of the contents and forms of each level of textual production (why and how was this written) should strive to connect them to contexts that are as precise as possible: socio-political, socio-religious, indeed, even economic.

The recent readjustment in Old Testament research on scriptural activity in the exilic and postexilic periods should remind us that just because of this abundance of literary output that research faces, it has raised more problems than it has obtained solid results. In order to avoid the dangers of arguing in a circle, which would consist of asking questions on the social context with regard to such or such theological positions of the text, settling them without sufficient external documentation, and using them as hypotheses for interpretation, it seems to us indispensable to broaden sufficiently the historical, archaeological and epigraphic documentation. Finally, these texts scarcely concern more than one ethno-cultural community of Achaemenid Transeuphratene. It certainly speaks of that community over and over again, but

Vulgate). It too seems to belong to the Persian period or to the beginning of the Hellenistic period.

8.   We can discern as well in some of these books (Ruth, Esther) as in Jonah and the story of Joseph (Gen. 37–50), a polemical current in the midst of the Diaspora, expressing a liberal theological view facing a closed orthodoxy, a current which subsequently persisted strongly enough to avoid seeing its witnesses excluded from the canon of Scripture.

only touches very slightly on the neighbouring populations: it is therefore impossible, for now, to appreciate the common points and the specificities of each of them through the texts of the Old Testament alone.

Chapter 8

POTTERY AND TRADE

Despite the fact that at present there are too few excavations on the sites
of Achaemenid Transeuphratene, as well as few reports on excavations,
the archaeological material is rich enough to be exploited, that is to
say, sorted out, classified, evaluated chronologically, qualitatively and
quantitatively, in order to establish some functional and historical
meanings. Since this material would be made up of objects of varied
and unequally represented categories, it would necessitate an experimental approach that was many-sided, flexible, progressive and continuously readjusted as a result of new developments in documentation
and new questions raised by historians. When we consider the most
complete excavation reports, the largest category of objects consists of
pottery, a category which archaeologists, trained to be specialists in
making typological and chronological comparisons, most readily use
as an indicator of economic and cultural relations, without, however,
detecting what resulted from trade and without being in a position to
reconstruct the latter.

A distinction must be made between pottery for local use, which was
not exported, and pottery which was exported, because of its intrinsic
value or because of its contents, to places more or less distant from
the centres of production, and which alone is significant for studying
trade. Now, the main difficulty that we come up against, especially in
Transeuphratene, is the localization of the production centres which
could make it possible to identify the two types of pottery, namely,
that whose use was strictly local and that which had a regional or a
little wider extension. There is particular uncertainty about the origin
of three types of containers widely attested along the coasts of the
Eastern Mediterranean and probably used for the maritime transport
of wine and oil: jars with straight shoulders, jars with basket handles,
and some amphorae with widened or egg-shaped bellies which could be

the prototypes of Hellenistic Rhodian amphorae. With regard to them, E. Stern proposed a hypothesis which would have to be verified: 'If the majority of these jars and amphorae were indeed locally manufactured their origins are to be sought in the East Greek islands and in Cyprus' (Stern 1982: 232). Likewise, the crude and very open bowls, incorrectly called 'mortars', which are especially found in the sites of the Eastern Mediterranean, always remain enigmatic. Archaeometric studies on the origin of ceramics, based on the analysis of its elements, attempt to establish the correspondences between the isotopic composition of a particular group and that of a reference material. While assuming that the analytical techniques (by neutronic activation in particular) and the methods of sampling would be perfectly controlled, studies of macro-origin* still come up against a major handicap, namely, the inadequacy of reference data, even for periods that are studied the most, like the Roman period. In our case, study of micro-provenance*, which is much less developed, seems the most useful for the ceramic documentation of the Persian period. An example drawn from the Yoqneam regional project, which covered the Western half of the plain of Jezreel in Palestine, can give an idea of some provisional, but significant results, to which these studies lead (Ben-Tor and Portugali 1987: 224-35). The analysis has focused on a certain type of Iron Age bowl, therefore relatively close to our period; they are cooking pots with a potter's mark (of the artisan and not the workshop), coming from several sites in this region. This study has made it possible to envisage the existence of an intraregional market for local ceramics, functioning on two levels: on one level, veritable monopolies in production, exchange, utilization within each of the geographical units (town, surrounded by villages); on the other, reciprocal but limited exchanges between different units at the level of the region as a whole.

A second problem to resolve is that of the dating of the bowls so as to be able to contextualize the exchanges, instead of situating them within a long period of time, which remains inevitable, however, when the shapes were in use over a long period. For the time being, the corpuses and the typological studies must be increased in order to give a starting point for ceramography, which in this case is still in its infancy. The Yoqneam regional project contributes to it, for example, by showing the statistical (in a percentage) and comparative value of a typology essentially established according to bowl rims (Ben-Tor and Portugali 1987: 139-223). In the same way, the typologies of the Iron

Age and the Persian period, established on the same principle at Tell
Deir 'Alla, in the Jordan Valley, and at Sarafand (Sarepta), on the
Phoenician coast, show first and foremost their regional range and
indicate under what problematic the research on their chronological
and socio-economic significance should be considered (Franken 1969:
102-249; Anderson 1975, 1987 [with bibliography]). For the Persian
period, it still all has to be done, if we wish to get beyond the stage of
'skeleton' corpuses of too little significance.

On the other hand, the imported Greek pottery on the Levantine
coasts is in general more easily identified and is often dated with relative precision. A certain number of difficulties remain, however: thus,
the production centres of the 'East Greek pottery' have still not been
clearly identified, it is not always easy to distinguish imported Greek
pottery from its local imitations, and it seems dangerous to be too precise in the datings, since the production of a master of a Greek workshop can be merged with that of his pupils and successors. That said,
well identified and dated Greek pottery is sufficiently abundant in some
regions of Transeuphratene that a historical interpretation can be undertaken, using when possible the quantitative study of archaeological
data in a critical manner (Elayi 1988: 11-12). As J.P. Morel with good
reason wrote: 'If the notion of dispersion is fundamental in the study
of traffic, that of quantity is as much and more, in regard to production and trade' (Morel 1983: 560).

This is why there has been a study recently, in typological and quantitative, chronological and spatial terms, of the Greek pottery, using
Greek in the broad sense of that term, imported into Phoenicia in the
Persian period, and why its relative significance in the reconstruction
of commercial flows has been recognized (Elayi 1988: 24-28, maps 2-
19, tables I-II). The role of Phoenicia in importing Greek vases into
Syria-Palestine in the Persian period has been known for a long time,
but the study of the chronological and spatial distribution of these vases,
by type and by area of origin, has made it possible to recognize the
general tendencies and a historically significant global evolution.

It thus appears that the dynamism of Attic production and export-
ing, which showed itself in the whole Mediterranean basin beginning
in the middle of the fifth century, especially with red-figured ware,
was strongly felt in central Phoenicia, but also in North and South
Phoenicia, as far as the interior of Palestine, in particular at Samaria
and Shechem. The pre-Hellenistic period (fourth century) had again

been marked by a new increase in imports, in the form of vases with black glaze and designs, in central Phoenicia, while they were reduced, on the contrary, in South Phoenicia, less, however, in the interior of the country than on the coast. The Attic pottery predominated in the volume of western imports in Phoenicia during the two last phases of the Persian period, and in this way took the place that the imports from Cyprus and 'Greece of the East' had occupied during the first phase (up to the beginning of the fifth century).

This general evolution seems to correspond, in a certain way, to the general political situation in the basin of the Eastern Mediterranean where the maritime hegemony of Athens asserted itself from the middle of the fifth century. It is generally admitted that this hegemony expressed itself in a commercial dynamism, which was at first in competition with the Phoenician cities' own dynamism, but combined with it in the end. It definitely seems, as a matter of fact, that the Phoenician commercial dynamism, especially the Sidonian, so clear in central Phoenicia during the fourth century, would have just followed the example of Athens. This dynamism seems to have been linked to a parallel political programme, initiated as early as the fifth century, and probably to a cultural tendency of assimilating Greek techniques, models, tastes, if not morals, inspired by Hellenic urban life. We know that the Persian chancellery, more preoccupied with keeping Egypt and its important resources within the bounds of the Empire than worried about the aggressiveness of Athens, saw a certain advantage in the development of commercial exchanges that enriched the Phoenician cities, and so did not interfere.

The analysis of the importations of Greek pottery into Phoenicia shows, on the one hand, that we cannot hope, in the present state of the published archaeological documentation, to find satisfactory responses to all the questions that trading raises, but, on the other hand, that the trends disengaged in a purely quantitative way, even out of so feeble a base, represent a methodological enrichment, since it makes it possible, through a more systematic approach, to renew the processing of the archaeological documentation in terms of trading, and through that, the questions of general history.

We will guard against improper interpretations of the importation of Greek pottery on the Levantine coast, proposed for a long time by certain archaeologists and now being progressively abandoned. Thus, it has been believed for a long time that the Greek pottery discovered on Near Eastern sites was used exclusively by Greeks and attests their presence in the region, from which it follows that there were Greeks almost everywhere. For some, only the Greeks would have been able to use the wine vessels since the 'Orientals' did not drink wine (see, for example, Akurgal 1966: 161-62), an anachronistic argument which is hardly put forward any longer, since the production and consumption of wine in the region in Antiquity are well known today through numerous textual and epigraphical sources (Delcor 1974; Rosen 1986–87 [with bibliography]). It has also been stated that the local imitation of Greek vases attests the presence of Greek potters, which in itself is absurd (Woolley 1938: 25-26; Boardman 1959: 163-69). As the points of view evolved, the next thought was that the discovery of Greek pottery on a site no longer attested the presence of Greek residents, but the fact that the transporters were Greeks. Thus, W.F. Albright relied on pottery to write: 'There can be no doubt that the coastal cities of Palestine and Phoenicia were leavened with Greek merchants and craftsmen at least a century before Alexander's definitive triumph' (Albright 1938: 34-35). We know today that the main trade was most often carried out through intermediaries and that as a general rule those who produced were not those who traded; according to ancient authors, the Phoenician merchants sold Greek, especially Attic, ceramics (Pseudo-Skylax 112; Pseudo-Aristotle, *On Marvellous Things Heard* 135 etc.): the traditional idea of 'Greek Trading Posts' in the Eastern Mediterranean area needs to be entirely revised today (Elayi 1987a).

On the other hand, we have too often extrapolated from pottery alone in studying trade. It is true that we are obliged to put ourselves in the perspective of what C. Renfrew called a 'pragmatic approach', that is, one based on the available material (Renfrew 1977: 1). Without going so far as to exclude pottery almost entirely in the study of trade (as some do: Mele 1979; Musti 1981), we must be aware of the restricted place it occupied. Thus, when Plutarch refers to the activities of the Athenians in the time of Pericles, he does not even mention the pottery, which was, however, extensively exported to the Mediterranean world at the time (Plutarch, *Pericles* 12.6); other cities that did not export pottery, such as Aegina, were no less rich trading cities. In fact, pottery should be considered, to use a phrase of D. Ridgway, as an 'inevitable side-effect' of a much more extensive commerce (Ridgway 1973: 23). If we consider, for example, the volume of trade between the Phoenician and Greek cities, it evidently cannot be determined globally because only non-perishable Greek products imported into Phoenicia remain for us and they could not have constituted the main volume of these exchanges, and because the perishable goods, coming from breeding and mostly from agricultural activities, by definition leave no traces.[1] Non-perishable products cannot in any case inform us about the general level of trade, or about its fluctuations, for there was not necessarily a balance between imports and exports, since the notion of a commercial balance was unknown at that time; in any case, the exchange value of the products is not available to us.

Furthermore, it must not be forgotten that the trade did not always take place directly between two cities; long-distance transportation in particular could not be carried out object by object and their provisional gathering together was necessary before distributing them again, along a capillary network. The interruption of trade in non-perishable products or its absence did not necessarily imply the absence of trading between two cities. Let us mention in passing that this trade was not, by itself, representative of the whole economy, but only of a kind of relatively well worked out commercial activity among other human activities, surpassing in particular the phase of family, village and even tribal self-subsistence of the agro-pastoral milieus of the Near East. Finally, from the study of trade we will exclude products levied, through authoritative channels, by an established political power whose centre of consumption and processing was situated at a certain distance from the places of production, or as spoils of war by an enemy army in the field: in both cases, the objects taken could go far, without strictly speaking becoming part of commercial networks.

As far as Phoenicia is concerned, the documentation from the Persian period, without being as rich as one might wish, is certainly not negligible. The classical sources mention the risks of sea-borne trade, its protection (Thucydides, *History of the Peloponnesian War* 2.69) and its regulation (Herodotus 2.179), the laying out of the royal route from

---

1. However, some products from the harvest or from processing, generally considered perishable, are of a kind that would leave more or less obvious traces, even outside of arid areas—namely, charred nuts, woods and grains relatively easy to recognize, and physico-chemical traces of organic remains (wine, oil), degraded but detectable in the laboratory. For all the more reason in arid areas, for example in those of Syria-Palestine, the remains of vegetable and animal fibres (cloth and mats), and of leather may be discovered: see Yadin 1966: 56-57, 196-97.

Sardis to Susa (Herodotus 5.52-53), an asset favourable to trade by
land, and the protection of caravan trade by fortified posts. On the
other hand, they are silent about the system of financing for commercial operations: did the Phoenicians make use of the advantages offered
by banks? We are not surprised at the tax privileges granted the merchants of Sidon by the city of Athens, when several texts of the time
confirm the role of the Phoenicians, principally as intermediaries, in
sea-going trade (see above, p. 114). The differences in the textual
sources leave many questions in abeyance, which we must try to resolve by archaeology.

The archaeological discoveries, of which pottery represents only one
aspect, provide supplementary pieces of information on the whole economic sphere, which makes it possible to situate trade in an appro-
priate framework. From the distribution area of each class of material,
established on the basis of exact types, we distinguish different types
of trade according to their range of contact. There existed in the first
place short-range trade, centred around the production of things to
use rather than on those for trade: the local and provincial markets were
circulation centres more or less closed in on themselves, taking up an
important part of local and provincial production for distribution on
the same scale. Side by side with this basically internal trade, there
existed medium-distance and long-distance trading, ranging up to 60
miles or more, whether caravan, river or sea trade, whose commodi-
ties for exchange are relatively easily spotted because of their 'exotic'
character, that is to say, differing from the local products. The products transported to a distance by land, particularly by caravan, were
generally non-perishable goods, especially processed or manufactured
products. On the other hand, it was especially products that were heavy
or supplied in large quantities, raw materials or products that had only
been processed in a preliminary way, whether perishable or non-perishable goods, which warranted sizeable shiploads going by sea: blocks
of marble, loads of wood, ingots of various metals, grains, oil and wine
in jars or amphorae; from these there remain today only the non-
perishable goods.

Thus we find in sites in the interior of Transeuphratene, and still
more in the Persian period than earlier, imported handicrafts: amulets,
seals, scarabs and other objects carved from stone or ivory, terracotta
figurines, crockery of stone, silver or bronze, metal objects (tools,
weapons and coins) and finally pottery. In the coastal sites situated at

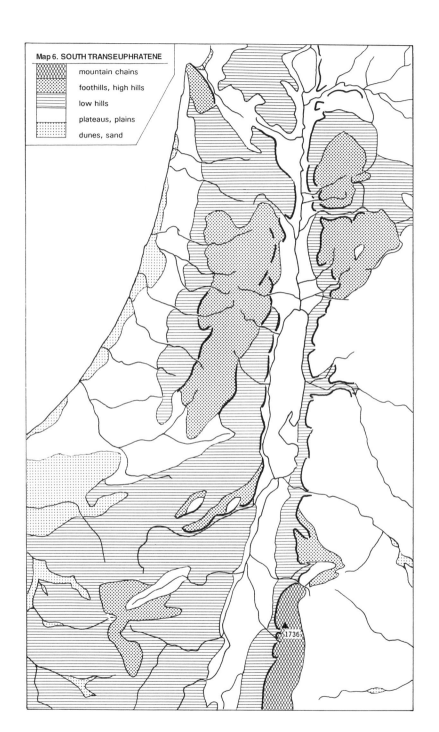

Map 6. SOUTH TRANSEUPHRATENE

- mountain chains
- foothills, high hills
- low hills
- plateaus, plains
- dunes, sand

1736

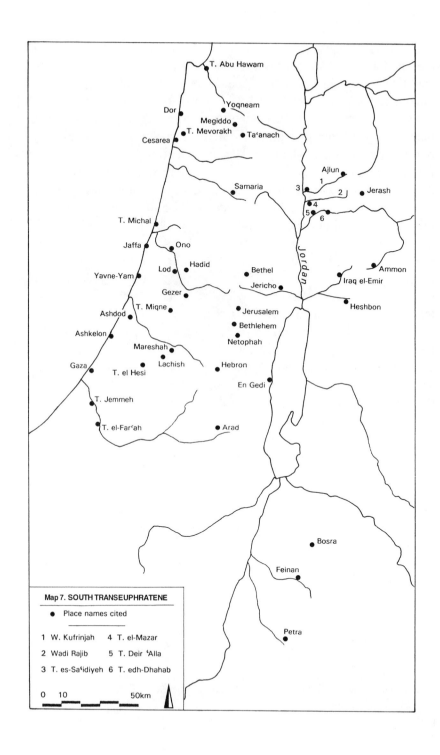

T. Abu Hawam

Yoqneam

Dor

Megiddo

T. Mevorakh

Ta'anach

Cesarea

Ajlun

1

3

Samaria

2

Jerash

4

5

6

T. Michal

Jaffa

Ono

Bethel

Ammon

Lod

Hadid

Iraq el-Emir

Yavne-Yam

Jericho

Gezer

Heshbon

Ashdod

T. Miqne

Jerusalem

Ashkelon

Bethlehem

Netophah

Mareshah

Gaza

Lachish

Hebron

T. el Hesi

En Gedi

T. Jemmeh

T. el-Far'ah

Arad

Jordan

Bosra

Feinan

Petra

Map 7. SOUTH TRANSEUPHRATENE

● Place names cited

1 W. Kufrinjah 4 T. el-Mazar

2 Wadi Rajib 5 T. Deir 'Alla

3 T. es-Sa'idiyeh 6 T. edh-Dhahab

0 10 50km

the junction of land and sea routes, or near these centres, we find on the one hand objects by artisans from distant lands and these objects could have come just as well by boat as on beasts of burden, and on the other hand marble, metals, jars and amphorae for transport.[2] It is especially on these sites that imported pottery occupies an important place in the material coming from the excavations; it shows the purpose of buildings where we find it sorted out by categories: jars for storage and for transport in the warehouses, splendid plates and dishes in the houses and residential quarters of the rich classes and/ or merchants. There is a renewed interest today in the metal trade, since the 'metallurgical purpose', as it is called by Ridgway (Ridgway 1973: 28), is starting to gain ground. It is being realized especially that the mines, considered too small today to be operated, were operating in Antiquity and had even sustained important trading trends: mineral archaeology, a quickly expanding development, certainly has much to teach us. The problem of supplying metals came up with special urgency for the Phoenician cities, which possessed few mineral resources, outside of iron, in their territories. We expect much for example from the exploration which is just beginning of the mines of Thasos, especially the 'Phoenician mines', which existed at least from the time of Herodotus and perhaps continued to be operated by the Phoenicians in the Persian period (see Elayi 1988: 72-73, 96).

Pottery is therefore just one indicator among others, and this category of imported and exported products can only indicate by itself the points of departure or the stops along the way and the arrival points, in which case we do not know if these were the real termini: the question comes up in particular with regard to warehouses. The trade routes still largely elude us at this stage of the documentation. To rediscover them, it is important to connect among themselves the various categories of archaeological documentation and integrate them, one to one and globally, into the geographical and historical approach to the economic phenomena and processes which they bring up. This comes down to taking apart and putting together again the internal and external pieces and functions of a system that is by nature complex, and of which many intermediate pieces and functions are lacking. Such an

---

2.   We should note that on the mountainous routes over bumpy ground, the containers for transport on beasts of burden were perhaps not fragile and inflexible pottery, but more flexible and sturdy enough containers such as goatskins: Mele 1979: 65; Maggiani 1972.

objective will certainly never be achieved perfectly, but if the imagination always inevitably comes along to make up for the deficiencies, the jigsaw puzzle of archaeological documents will only be able to be pieced together organically by an always more thorough and better established knowledge of the functional relations between the objects and structures of all categories, the economic and other processes, and of the more or less specialized human groups who were involved in them.

If the products which have undergone primary processing are not sufficient by themselves to vouch for the existence of a commercial activity, archaeology finds in particular installations, such as workshops, evidence of craft work which, going beyond the bounds of a purely family or local economy, must have been oriented towards trade. We can thus consider as starting points for trade movements the following: for the period of Iron IIC (seventh century), the olive oil installations of Tel Miqne (Ekron of the Philistines) in the Judaean Shephelah (Dothan and Gitin 1987: 208-10; 1988: 234-36); for the Persian period, the numerous wine and oil presses of the farms dispersed throughout the hills of the province of Samaria (Zertal 1990; Finkelstein 1981; Dar 1986), or again the site of copper metallurgy at Feinan in South Jordan, where the important volume of slag indicates an intense and centralized activity (Homes-Fredericq and Hennessy 1989: II, 225-27 [with bibliography]). Some storage installations (Gubel 1990) could have been used to handle these trade movements, but we must wait for other pieces of evidence to justify such an interpretation. Some caravan stations in the interior or in the coastal regions, recognizable from their architectural structure and the remains of their equipment, could theoretically provide supplementary information on the nature of commercial currents (Reich 1984), but we are still far from having available an interpretable documentation in this sense or in any other.

The combination of ancient sources and the archaeological discoveries introduces us to some trading posts connecting the Phoenician cities to the Greek cities and to some of their extensions towards the hinterland: the Cypriot cities, Phaselis on the coast of Lycia, Naukratis on the Canopic branch of the Nile and Tell el-Kheleifeh at the head of the Gulf of Aqabah. There has been much discussion on the exact nature of the great trading junctions. The main difficulty seems to come from the fact that the 'port of trade' which, moreover, was not necessarily a port, but a contact point and a gateway, obligatory or preferred, between two ecological environments, could have been of different types: a port within the limits of a city (Piraeus); an emporium* spread out to the city limits (Byzantium); a port situated in one country but dependent on another country (an Assyrian *Karum** in the city of Arwad) (Elayi 1988: 76-77 [with bibliography]); or whatever.

With regard to the commercial crossroads in the interior of Transeuphratene, they do not provide as much valuable information as we would have the right to expect. The sites in the cities, whether large or small, remain practically silent for three principal reasons: in certain cases (Aleppo, Damascus, Jerusalem), we cannot reach levels of the Persian period over extensive enough areas because they have been permanently inhabited from antiquity up to our own times; other sites, such as Tell Nebi Mend, Megiddo, Samaria and Gezer, were excavated too early, without an adequate method or technique; in the case of others, such as Tell Mardikh, Neirab, Tell Deinit or Hazor, their urban nature is not certain.

There remain the port sites of the Mediterranean coast that profited from a double commercial traffic, caravan and maritime at the same time. This latter, essentially operating under the form of coastal navigation, duplicated to some extent the caravan traffic for transporting bulky and heavy products. The warehouses of the actual port installations still remain unknown to us; these are the quarters of the merchants which are capable of illustrating the departures and arrivals of the land- and sea-traffic flows. Again, it would have been necessary to have excavated all the warehouses and quarters of a given port site, to have preserved and published all the material, in order to obtain a balanced picture of the nature and relative quantities of the non-perishable trade at that site. It is only on this condition that we will be able to reach an overall view on one of the centres of great maritime commerce.

To sum up, archaeological exploration constitutes a rich source of information on various aspects of the economy of the region. The ill-considered use that has been made of the diffusion of pottery to define trading patterns can be considerably improved, if this category of objects is integrated with others, and if trade is not separated from the totality of economic activities. In the framework of those activities that it is advisable to apprehend on different spatial and temporal scales, the installations of all kinds, with their furnishings *in situ* which archaeology turns up, constitute the indispensable basis for the study of the activities of production, storage, usage, consumption and exchange, which can help define the spatial structures of trade movements, their nature and their volume.

Chapter 9

NUMISMATICS AND ECONOMIC HISTORY

The economic history of Transeuphratene in the Persian period to a
great extent still has to be written, not for lack of information, but
rather for lack of interest on the part of specialists on the ancient Near
East. Certainly, we do not have at our disposal a sizeable mass like the
Mesopotamian economic tablets, but the analysis and collation of pieces
of information of various origins, in particular the archaeological data
of which we have just spoken, should make it possible to deal with the
main problems of economic history that come up in our field, namely,
economic resources of different regions, relations between economy
and society, the disappearance of the palatial economy with the resulting benefit to the economy of the city, possible persistence of agro-
pastoral economies functioning in an independent way, the place of the
various types of economies in the framework of the imperial Persian
economy,[1] arrival of the monetary phenomenon, and so on. It is this
last aspect, so extremely important and of exemplary value, that we
deal with here, along with its economic implications (Elayi 1989: 197-
233 [with bibliography]).

Transeuphratene, and more particularly the Phoenician cities, represent one of the regions of the ancient world where monetary problems take on the greatest complexity, at least as far as the Persian
period is concerned (Elayi 1988: 39-60). In fact, it is possible to find
in the same city, successively or simultaneously, Persian coins, imported Greek coins, counterfeits of these coins, Cypriot coins, satrapal
coins, local coins, coins of neighbouring cities or provinces, whether
Phoenician or 'Philisto-Arabian', each category, moreover, bearing
countermarks.

1.   Studies on these aspects of economic life have been infrequent: Bondi 1978;
Elayi 1990c.

Now, all these coins of such a varied nature have been for a long time—and still are today—classified among Greek coins, and studied, because of this fact, by numismatists with Hellenic training; sometimes however, Phoenician coins have interested numismatists with training in biblical studies, from a Bible-centred perspective. In the first case, they only considered the typology while abstracting from the inscription, while in the second, they interpreted the typology and the inscription in reference to the Bible: in both cases, knowledge of the specific context of the coinage, indispensable for a serious numismatic study, was non-existent or insufficient. Very fortunately, numismatics, which for a long time has remained one of the most fossilized social sciences, is on the way to transforming itself, by especially opening itself up to economic history and by using the help of new scientific disciplines. The two recently organized Colloquia relating to our field of research illustrate this significant change: that of Louvain-la-Neuve in 1987 on *La Numismatique et l'histoire économique dans le monde phénico-punique*, where specialists in Phoenician–Punic studies took part, and that of Bordeaux in 1989 on *L'or perse et l'histoire grecque*, where Iranist and Hellenist scholars dialogued.

The traditional criteria for numismatic analysis should not be abandoned, however, but they need to be renewed and completed by the new techniques and problematics. Typological study is always indispensable for the identification and classifying of the coins, but interpretation of the symbols should not lead to extrapolations based on deficient knowledge of the local context: thus, the soldiers represented on the deck of the galley on the shekels of Byblos are not 'hoplites' as has long been thought, but marines of the naval force that Byblos had just built up again towards the end of the third quarter of the fifth century (Elayi 1984). The owl on the shekels of Tyre has nothing to do with the Athenian owl: its appearance is based on the Egyptian hieroglyphic sign, 'M', but is distinguished from it by the fact that it does not bear the flail alone, but the flail and the sceptre hooked; it is a matter, then, of a symbol to be interpreted on the strictly local level, whose meaning escapes us at present (Elayi 1983: 12-13). It is necessary to be attentive, too, to the location of the symbols and to ask why, for example, the leaping dolphin of the first series from Tyre became a secondary motif inscribed below in the series with the divinity on the sea horse. Despite its stylized character, the representation of the warship on the coins of Arwad, of Byblos and of Sidon, provide valuable details on the different techniques in the navies of each city and on the modifications brought in during the Persian period.

The usefulness of stylistic study has long been overvalued for establishing the classification of coins; it should be reduced to its true value, as just one way, among others, of doing a classification. Chronology cannot be based in any case on a supposed improvement in the style, since examples show that deterioration and improvement can follow one another indiscriminately: thus, the very first Phoenician issues have a very meticulous style.[2] Stylistic blunders are due to less skilled or hurried engravers and usually reflect either a speeding up of the pace of monetary production, which obliged them to work too fast, or the striking of an emergency coinage for which they could have been forced to hire less qualified engravers.

The study of die-linkings* constitutes one of the surest means to establish the classification of coins, but it is advisable to remain very prudent in coming to conclusions about similarities; in doubtful cases, the use of a comparator of dies* can eventually help in resolving the problem. As a general rule, it is better to refuse to take into account two dies as identical rather than consider as identical two different dies, which could lead to false die-linkings. The study of dies also makes it possible to evaluate the volume of the issues and the rate of monetary production, which is important from the point of view of economic history. For now, it is premature to speak of studies of dies in our field, since such studies can hardly take place outside the global perspective of a corpus and none, even a partial one, as yet exists on the coinage of Transeuphratene.[3]

The study of the weight of coins is always useful,[4] on condition that it focuses on a sufficient number of examples to be really significant; on the other hand, it should take into account modifications due to several different factors, such as the cleaning of coins. It is also a matter of knowing how to ask good questions, for example, on the identification of the monetary standard; on the reasons which could have

2.   For example, the first coins from Byblos, which have a crouched sphinx on the obverse and, on the reverse, a stylized double lotus.

3.   Some numismatists however use a partial method based on statistics in studying a part of the dies; see for example Hackens and Carcassonne 1983; Carter 1983; de Callataÿ 1984.

4.   It would be desirable to develop metrology, a discipline a little too frequently forgotten and still in its infancy in our field.

determined its choice and, eventually, its variations or its alteration; and on the relation between the standards used and the local weight systems. We will be especially sensitive to any reduction in the weight of the standard that could reflect financial difficulties of the issuing government or difficulties in getting supplies, insofar as the source of metals for minting was not available in the territory of the issuing government: how are we to interpret, for example, the lowering of the weight of the Sidonian double shekels* by almost two grams beginning with the reign of 'Abd'ashtart I (Elayi 1989: 213)? It would be tempting to link this phenomenon to the deterioration of the political situation in the region and more particularly in the city of Sidon, with its open opposition to the Persians; but as there was no silver mine in the territory of Sidon nor in its vicinity and since we do not know where the Sidonians got supplies at that time, there is no reason to hold the political situation rather than the source of supplies responsible. The analysis of the composition of the alloy of the coinage makes it possible finally to determine the variation in the fineness of the metal and to detect devaluations not visible in the weight: there again, everything still remains to be done.

The study of the techniques in manufacturing is generally worthwhile, since it can provide a lot of different pieces of information. Overstriking* is sometimes easy to detect and significant when it is systematic; thus, the series of Sidonian double shekels which can probably be attributed to Ba'alshillem had been partly overstruck on the preceding series with the galley in front of the fortifications (see, for example, Hill 1910: 143 n. 17-19). The very obvious overstriking, as in this case, could have conveyed the desire of the issuing government to show that it was using, while destroying them, the symbols of the preceding political authorities; but as this overstruck series immediately preceded the series with the reduced standard of 'Abd'ashtart I, the overstriking seems to be explained rather by a shortage of metal. There has been a study recently of the techniques used in the manufacture of Phoenician coins (Elayi 1991c): we now know that the engravers worked on the dies by directly engraving, using partial punches shaped like beads for the milled edges and the marks; they carried out preparatory work on the surface of the die by imprinting markers (rings and lines); they then engraved the decoration, next the inscription and finally the circle of the perimeter.

As we saw in Chapter 6, p. 88, the study of coin inscriptions is an essential element in the study of numismatics, and can provide valuable historical information besides. But we are still quite often at the preliminary stage in determining the readings, a prerequisite to any attempt at interpretation. Thus, the divisional coins of Tyre with the dolphin bear an inscription that scholars have interpreted in various ways without having solidly established the reading; they were content to work with quite poorly preserved examples, on which the form of the letters is not clear, or with examples whose motif is off centre and whose inscription falls outside the flan*. Before any attempt at interpretation, it is necessary first of all to gather together all its known examples and to determine the reading according to those examples whose inscription is well preserved and centred, namely, clear and complete (Elayi and Lemaire 1990).

The discoveries of coins can come up under different more or less significant forms.[5] Isolated coins are of little significance; nevertheless, a concentration of isolated coins in a limited area can provide valuable information on the circulation of coins in a region in a given period. The coins from excavations can also provide details on the currency in circulation in the excavated site, if they are studied in a critical way: thus, the numerous coins from the Persian period discovered on the site of Al-Mina, at the mouth of the Orontes, are almost all coins from Arwad; owing to this find, in addition to other indications, it has been possible to demonstrate that Al-Mina in this period was not a Greek trading post, as has always been thought, but a Phoenician town, most probably dependent on the neighbouring city of Arwad whose coinage they used (Robinson 1937; Woolley 1938; Elayi 1987a). The study of the distribution of monetary discoveries makes it possible for example to determine that Byblos coins hardly circulated outside the territory of that city, whereas the coins of Tyre and Sidon were very extensively exported to Palestine, and at the same time circulated in the Phoenician coastal possessions and in the hinterland, with varying functions (Elayi 1988: 67, 174, 200-202).

The study of hoards of coins is more or less fruitful according to the type of hoard under consideration; it could be a matter of accidental losses, something hastily buried, treasure hoards*, or deliberately abandoned treasures. The discovery at Al-Mina of a purse containing 54

---

5. On the different types of treasures and their interpretation, see Elayi and Elayi 1989: 217-24.

Phoenician coins of different denominations lets us see what an inhabitant of that city could have had in a money pouch for current purchases (Robinson 1937: 185-86); this type of treasure, which reflects well the circulation of coins at a given time, rarely represents a round sum, since the coins were not put aside for a specific purpose. Treasure hastily buried, at the time of a military, political or economic upheaval, represents the great majority of the treasures discovered in Transeuphratene; their owners were not able to return to recover them, either because they had been killed or taken prisoner, or because the location of the hiding place had become part of enemy territory, or simply because they had forgotten it. The coins which make up this type of hoard had been taken out of circulation at the time of the burying and as a result reflect too the circulation of currency at that moment; these treasures are often dated by the event that led to their being buried, especially when we find, in the same region, several of similar composition. Most of those in Transeuphratene were buried during the second third of the fourth century, a particularly troubled period, either at the time of the great wave of revolts of the satraps, or at the time of Alexander's conquest.[6]

The discoveries of gold Persian darics or silver Persian shekels are quite exceptional in the region: there is just one daric for all of Palestine, as far as we know, and it comes from the regular excavations of Sebaste (Samaria); besides, it could be later than Alexander's conquest; some Persian coins are found in the hoards from Hauran, Massyaf and Anti-Lebanon.[7] We cannot speak at any time of a cash economy in connection with the Persian coinage, which had above all a military, political and social function, and was rarely used in business dealings.

Greek coins had been extensively imported into Transeuphratene in the Persian period (Elayi 1988: 41-45). Alongside Athenian coins, which make up the vast majority of the imports, we find 'Thracian–Macedonian' coins in the sixth century and Cypriot coins all through the Persian period. Coins from other Greek cities are scarcely represented at all, which is not surprising since these coins were for local distribution. Although the Cypriot coins were for local distribution too, the fact that a certain number of them were found in the Phoenician cities could be explained by the geographical proximity and by the

6. See for example Thompson *et al.* 1973: nn. 1485-1506; Naster 1987.
7. Fulco and Zayadine 1981: 199; Thompson *et al.*: n. 1483; Kraay and Moorey 1968: 192 n. 90, 218 n. 83; Hurter and Pászthory 1984: 111-25 nn. 50-53.

bonds uniting some Cypriot cities to their mother-cities in Phoenicia. Almost all the imported Greek coins were tetradrachms; decadrachms, octadrachms, didrachms, drachmas, and the small denominations were quite exceptional. What was the function of these Greek coins imported in such great quantities? They did not have a monetary function, but circulated as valuable merchandise; in fact, in contrast to the treasures discovered in the Greek world, in the treasures of Greek coins of Transeuphratene there are often found pieces of gold or silver plate, some ingots and various scraps of molten or cut silver. Greek coins were probably valued insofar as they were considered a guarantee of a good standard*; several of them had received one or more blows of chisels, as if someone wanted to make sure that they were not counterfeit. The relative abundance of imported Greek coins in the Phoenician cities cannot be explained by the fact that their trade with Greek cities was more developed than that elsewhere, since, in some regions where trade was especially developed, at Syracuse for example, we find very few. The most logical explanation would be that the Phoenician cities deliberately accumulated Greek coins because they lacked silver; they were used as valuable merchandise in cities without their own coinage and were probably in part reused in the coinage of cities that bartered money; this was done by recasting rather than by overstriking, since the latter procedure would have left traces.

Imitations of Greek coins form a category abundantly represented beginning at the end of the fifth century, from Egypt to the Bactria, and in Transeuphratene almost all are 'pseudo-Athenian' (Elayi 1988: 45-48; Nicolet-Pierre 1986). In these coins, carefulness in imitation was essential, but there was a great deal of variety in how well this was achieved. When the copies are precise and engraved well, they are hard to distinguish from the originals, all the more so since they are sometimes mixed in with the originals, as in the treasures of Al-Mina. However, most often, the 'pseudo-Athenians' are distinguishable from their models by one or more details or by the addition of an inscription: thus, a 'pseudo-Athenian' tetradrachm from Al-Mina does not have the crescent on the reverse side; another, a perfect imitation, has on the reverse side two Phoenician letters whose meaning escapes us. The 'pseudo-Athenians' can also be distinguished by a particular style, that is occasionally clumsy. The authorities issuing these coins seem to have been most diverse, namely, satraps, cities or even a king of the

Persians, Artaxerxes III (in Egypt). The imitation gives them a different signification from that of civic coins: they were probably intended
to answer a demand for Athenian coins, which were scarce, at a time
in which there was a decline, especially in Phoenicia, in imports of the
coins with the owl on the reverse side. They were probably used then
for the same purpose as their models; as they were always of high value,
they could have served as ingots in important transactions, for example, in the commercial town of Al-Mina. They were probably intended
for local merchants rather than for Greek merchants; however, it
could be that the Greek merchants had also accepted the ingot money
in payment, in case of need. But, among other uses, the most probable
and the most frequent was seemingly for non-commercial payments,
of the mercenary type: this use is attested by the 'pseudo-Athenian'
coins issued by Tachos on the advice of Chabrias to pay the Greek mercenaries in Egypt, who particularly prized the Athenian tetradrachms
and the Persian darics. The coincidence between the development of
the use of Greek mercenaries in Transeuphratene and that of 'pseudo-
Athenian' coins is perhaps not by chance. On the other hand, the penetration of these coins even into Athens is significant: if they could have
been brought back occasionally by Greek merchants or Near Eastern
ones passing through, even though it would not be very probable that
silver would have been exported from an area where it was scarce, we
must think rather of Greek mercenaries employed in Transeuphratene
who brought back their wealth to Greece, or whose wealth was repatri-
ated after their death, as was the case with an Athenian mercenary
who died at Akko (Isaeus, *The Succession of Nicostratos* 76.7). The
Athenian decree of 375/374 confirms that these 'pseudo-Athenian'
coins were accepted by the Athenian State on condition that they were
genuine (Stroud 1974; Buttrey 1979).

Apart from the 'pseudo-Athenians', categories of coins that resemble
closely or distantly a Greek model and whose date and precise place of
issue are unknown are grouped together under various headings, such
as 'Philisto-Arabian'; they are not homogeneous, either from the iconographic point of view or from the point of view of the source issuing
them, or from the point of view of their destination. Their types, which
are extremely varied, stand in contrast with other monetary types,
which are quite stilted: they feature personages, real or mythical animals represented fully or partially (head or bust), objects (a flower or
a citadel, for example). Their distribution zone is essentially Palestine,

Egypt and Arabia; little has been found in central Phoenicia and Syria (Hierapolis/Manbog/Menbij). The dates of issue are very uncertain; they could be stretched out between the middle of the fifth century at the earliest and the end of the Persian period, with an important development in the fourth century. The authorities issuing these coins seem to have been as varied as those who issued the pseudo-Athenians, namely, rulers of local dynasties and official representatives of the Persian authority. The coins bearing the name 'Yehud' (Judaea) are the best known; they were struck in the fourth century at Jerusalem by the governor of Judaea; those which bear the name 'Hezekiah' were probably issued by the high priest with this name when he was governor of Judaea (Rappaport 1981). Only three other workshops have been identified up to now: Samaria, Ashdod and Gaza (with two issuing authorities: the city and the Arab dynasties) (Mildenberg 1990); the question of the existence or not of a coin workshop at Ascalon has not been resolved and there probably still remain several other workshops to be discovered.

Phoenician coins, it seems, began to be struck beginning in the middle of the fifth century, and perhaps even a little before, by the four workshops functioning in the Persian period: first Byblos, then Tyre, Sidon, and finally Arwad (Elayi 1988: 50-53 [with bibliography]; 1989b: 197-233 [with bibliography]). Everything took place as if, after the initiative taken by Byblos, three other cities, one after the other, followed that city's example. It could seem astonishing that Phoenician coinage had appeared with a delay of at least a century and a half after Greek coinage, even though these coins were well known in the region and Phoenician commercial activities were similar to Greek activities and all the technical conditions were present, namely, a good knowledge of metal working and practice in engraving on seals. The real reasons for the arrival of the currency phenomenon in the Phoenician cities still escapes us, but it should be noted that the context was different at Byblos, where the city at the height of its prosperity had constructed a fleet for itself, and in the three other Phoenician cities, the greatest part of their naval potential had been destroyed in successive wars, in the service of the Persians; in order to handle all the expenses in the reconstruction of the destroyed ships, these three cities could have tried to derive profit from the difference between the value of raw silver and the legal tender for coins. Striking coins represented as well a way for a city to affirm its independence or its autonomy;

because of successive defeats suffered since 480, the relations of Tyre, Sidon and Arwad with the Persian occupation had deteriorated and their public image had probably been tarnished: coinage constituted an excellent instrument for political propaganda.

Besides their eventual financial and political function, did Phoenician coins have a commercial purpose? Actually, the Phoenicians had not the least need of currency for their trading; in the important maritime commerce, use of money was uncalled for, since the ships were not able to return with holds loaded with just a cargo of coins: the merchants used a series of exchanges without any need of a real monetary component. The cities would not have taken on the further expense of minting coins just to be used as currency in local commerce. In land-based international commerce, the Phoenician coins of Tyre and Sidon were probably used as commodity exchange in the same way as Athenian coins, as is proved by their discovery in some hoards reduced to fragments or mixed in with pieces of raw silver. Shekels from Tyre and double shekels from Sidon, whose symbols guaranteed the quality of the silver, were exported to the regions bordering on their territory and in the fourth century seem to have been in competition with Athenian coins that arrived in the Near East in lesser quantities; it is not impossible that some 'Philisto-Arabian' coin workshops who were short of silver recast and re-used them.

If the commercial function was not clearly apparent in the first issues, the large Phoenician cities definitely seem to have experienced subsequently, in the fourth century, the emergence of the cash economy. In fact, all the conditions came together—namely, regular issues, sufficiently diversified denominations to allow for all types of transactions, particularly with the appearance of denominations in bronze, and finally an abundance of the quantities issued. With a cash economy already having begun, it would seem, in some Greek cities in the middle of the fifth century, it is possible that the Phoenicians had become aware of how useful Greek coins could be as a means of payment. Entrance into the cash economy amounted in the Phoenician cities to a real economic 'cultural change', a fundamental event, since by itself alone it implied profound modifications in the concept of trade. This change was not so much the fact that coins were issued, as that the users had to modify their commercial operations, with consequences that are still impossible to evaluate.

Alongside the considerable impact of this phenomenon, it is necessary, too, to emphasize its limits: it was, in fact, limited to trade within Phoenician cities, and was non-existent in the case of external trade; on the other hand, even in internal commerce, the ancient systems of exchange survived and were not ready to disappear completely: barter systems, metal weighed and counted. This complicated situation seems quite similar to that of internal Athenian commerce in the fifth century, as can be seen in a passage of *The Acharnians* of Aristophanes; the Athenian Dicaiopolis asked a merchant from Thebes, who had come to sell some goods in the Athenian market, what trading system he used: 'Let's see now; what price do you charge? Or instead are you taking back there (to Thebes) other goods from here?' (Aristophanes, *The Acharnians* 898-99; see also 813-14, 830-32, 900-904).

What took place in the rest of Transeuphratene? There definitely were regions which only used weighed or counted metals for money, and did not themselves have monetary workshops. Philisto-Arabian coinage is insufficiently known for us to be able to decide whether their issuing authorities had accomodated themselves to a cash economy; at the present stage of the documentation, we do not get the impression that they would have issued quite varied denominations, but perhaps they had not yet given any attention to smaller denominations. In any case, the question does not even come up, since in the first place it is a matter of identifying the workshops; these workshops seem to have been numerous in the fourth century, highlighting the emergence of local governments, whether dynastic or priestly, and of provincial or district governors, perhaps to distance themselves in relation to the central Achaemenid government. We can measure at present the distance that still has to be travelled in order to understand properly the appearance of this monetary phenomenon which had a limited adherence to a cash economy, and which represents, however, only one of the least badly known aspects of the economic history of Transeuphratene.

Chapter 10

A SPECIFIC HISTORICAL SOCIOLOGY

For some years the social sciences have begun to be included in studies
on the ancient Near East, principally in the Old Testament area. Even
though we can already discover some isolated sociological remarks
in the work of Flavius Josephus, biblical scholars have traditionally
centred their interest on philological, literary and historical questions
and hardly ever preoccupied themselves with the structures and social
evolution of the milieu suggested by the texts which they studied.
W. Robertson Smith in Great Britain, L. Wallis in the United States
and especially M. Weber in Germany were the main ones who began to
show interest in a sociological approach to the biblical world (Robertson Smith 1889; Wallis 1912; Weber 1952; see also de Robert 1984:
403-404). If sociology is a quite recent discipline, historical sociology
has scarcely taken its first halting steps. This situation is reflected in
the terminology used, which is characterized by a proliferation of
terms whose meanings differ from one language to another and sometimes even from one school to another. Without going into detail, we
note that in France they use the term 'anthropology', but still more
often the term 'sociology', which appeared for the first time in 1837
in the fourth volume of *Cours de philosophie positive* of Auguste Comte
(see, on this subject, Szacki 1979: 185), and designated the study of
social organization, including beliefs, customs and folklore (Gurvitch
1968: 19-27). It is the equivalent of the English term 'anthropology',
since 'sociology' in English has a more restricted sense (Herion 1987:
44-46). It is also the equivalent of German 'Soziologie' and 'Ethnologie', with the term 'Anthropologie' designating the theories and speculation on the nature and existence of humans. In the United States,
anthropology is divided into physical anthropology and cultural anthropology, with the latter category including social and structural anthropology, ethnology, ethnography and archaeology (Wilson 1984: 17).

In the field of ancient Near Eastern studies, several avenues of research have begun to be explored from a sociological perspective, but with uneven results (Lang 1983); the ancient periods have been favoured up to the present, to the detriment of the later ones, particularly the Persian period. Recent works indicate a new interest in the social reality of biblical history, either utilizing sociological concepts in a more descriptive rather than analytical manner, or in applying specific sociological theories to particular problems: it is thus that they have proposed some reconstructions of pre-monarchic Israelite history, starting from operative models from modern sociological studies.[1] Works of this type are criticized, sometimes with reason, for being based on certain presuppositions—namely, positivism, when sociology is ranked among the number of empirical methods that can prove useful to the historian; reductionism, when an attempt is made to reduce the complex to the simple; determinism, when it is thought that human values and actions are determined by variables; and relativism, when everything is connected to socio-cultural circumstances (Herion 1987: 54-56). The search for methodological models that are always operational does not appear to be borne out in our field; in the same way, the four categories of W.G. Runciman—reportage, explanation, description, evaluation—seem poorly adapted to literary analysis of the Old Testament.[2]

When C. Lévi-Strauss created structural anthropology in 1958 to describe social facts in terms of an underlying system as in linguistics (Lévi-Strauss 1958; 1973), he did not wish to apply it to the Old Testament which he considered too fragmentary and subject to theological reinterpretation. However, his method has been utilized on several occasions in the Old Testament field, for example, for the study of the binary structure of the representation north–south, for the structure of myths, or again for the structure of kinship (Rogerson 1974; Pocock 1975; Polzin 1977; Rogerson 1978: 102-19; Wilson 1984: 22-23). Outside of this field, we draw attention to some attempts at research on the structures of kinship—namely, that of C. Herrenschmidt on the Achaemenid Persians, based on royal inscriptions and the classical sources, which have proved to be promising (Herrenschmidt 1987); and that of D. Henige on the kings of Sidon in the Persian period from Sidonian

---

1. Sasson 1981; for an overall view on the research in this field, see Mayes 1988.
2. Runciman 1983; see the criticism of Rogerson (1983: 256).

inscriptions and the Classical sources, which do not seem to us to provide in this respect sufficient data (Henige 1986: 57-60).

The introduction of anthropology into the archaeology of Transeuphratene has been shown to be profitable by R. Poppa's publication on the necropolis of Kamid el-Loz (Poppa 1978; see on this subject, Sapin 1989). He carefully analyses the material details that draw attention to particular funerary rites—namely, the lying in state of the body, the burial, the offerings and accessories, including apparel and ornaments. He particularly documents the low proportion of infants' tombs in comparison with those for adults, which does not correspond to the proportions of a normal ancient population, experiencing a priori a high infant mortality rate; on the other hand, we have here only the tombs of the richest families, and not the tombs of warriors, and just one tomb of an artisan. In this necropolis, the whole population of the ethnic unit was not represented and therefore had other places for burying their dead. As a hypothesis to be verified, we could imagine a form of semi-nomadism with two variables: in most favourable periods of political calm and sufficient rainfall, the group would split into two groups, leaving one group in place which would carry on with agriculture, while the other group would be leading the flocks either toward the Syrian steppe, or toward the Lebanese mountains, or again toward the two successively; in periods unfavourable in every way, the whole group would find itself leading a nomadic existence on the steppe. The anthropological perspective of R. Poppa spills over into the question of social, economic, political and cultural structures and systems of the Lebanese Beqaa in the Persian period.

Physical anthropology, when well understood, can bring a non-negligible complement to the interpretation of archaeological material. Thus, the study of human bones of the first and third centuries of our era at Ein Gedi and Yavne Yam has made it possible to determine the cause of death and the type of weapon used (Rak, *et al.* 1976: 55-58). The study of the mummy of King Tabnit of Sidon has shown that he died in the prime of life—which is confirmed by the inscriptions; his death followed a long illness (smallpox), since he had not dyed his hair with henna for some time (Jidejian 1971: 117). Mention must be made, too, of the works of the American team of L. Stager, and H. Bénichou-Safar which, although dealing with the bones of children in the always mysterious tophets* of Carthage, has value as a model (Bénichou-Safar 1987 [with bibliography]): the study of the cremation of bones with the help of legal-medical experts provides information on the position of the bodies, the way they were clothed, the nature of the faggots and of the place where the cremation took place.

We will not elaborate on religious anthropology, defined as the knowledge or science of religious people, which has inspired many studies in biblical theology (Dupront 1974), nor on the anthropological approach to iconography, which brings a new dimension, particularly to the study of innumerable Near Eastern cylinder seals (see, for example, Durand 1984). The application of cultural anthropology to research on isolates* on the ground is worth pursuing too, as long as it is rigorous and offers directly observable parallels, likely to clarify ancient data. It is especially in the sphere of the permanence of techniques that research has been carried out, particularly in Iraq and Saudi Arabia (Chelhod *et al.* 1984; Balfet 1980; Postgate 1980): thus, the cooperation of an archaeologist and an ethnologist has made it possible to know more about those who worked in clay; in the same way, starting from the study of modern isolates, a better understanding has been attained of the use of reeds and palms in Mesopotamian construction techniques. As for Transeuphratene, at 'Aqura, a village in the Lebanese mountains, isolated in the upper valley of Nahr Ibrahim, evidence has been uncovered of the permanence of the technique of Phoenician pillar walls, principally developed in the Persian period (Elayi 1986).

Historical sociology has proved particularly useful in the study of cultural contacts and this is especially interesting in connection with a region where so many different populations coexisted, mingled or interpenetrated. We will try to illustrate, from an example in this particular area, what could be the foundation of historical sociology. Among the positive aspects of this new discipline, it is acknowledged that in general it sensitizes historians to the social dimension of their research, something neglected up until now; it gives rise to new questions; it opens up access to a genuine interdisciplinarity; and it can result in a contribution to modern sociology (Wilson 1984: 28-29, 53, 81-83; de Robert 1984; Herion 1986: 22-26). The main criticisms directed at it are the following: given the number of schools, the divergence in methods and the frequency of disagreements, it is difficult to make a judicious methodological choice; the value of the published works is very uneven and a number of them are ideologically biased; generalization is dangerous; modern models are difficult to apply to ancient societies, whose context is always different, and they cannot be tested back there; finally,

Beyond the River

historical sociology is sometimes done to the detriment of traditional historical methods and never allows for a validation of a historical reconstruction (Rodd 1981; Rogerson 1983: 249, 256; Wilson 1984: 53; Herion 1987, 57; Jobling 1987). Those most pessimistic conclude that a historical sociology is impossible (Rodd 1981: 105), but most often the following conditions and limits are placed on its use: it is indispensable in the first place to know modern sociology and its models well, in order to be able to apply them to ancient societies; it is necessary to be attentive to the differences, and not only to the resemblances; finally, this new approach can only be considered something supplementary and its results should be rigorously controlled by other data, particularly textual data (Wilson 1984: 28-29, 53; Herion 1986: 23-24).

We will take the example, in a recent study, of the cultural contacts between the Phoenicians and Greeks in the Persian period (Elayi 1988). If historical sociology is still in its infancy, the study of the interpenetration of societies involves one of its most neglected aspects. The historians who were the first to practise it have never gone beyond the descriptive phase. When sociologists did become interested in it, they especially understood it on the level of primitive mentalities or imagined it as a social pathology of the clash between our Western civilization and the traditional civilizations; because of this, historians of Antiquity have long been hesitant to adopt this new avenue of research (see, for example, Van Effenterre 1965; Momigliano 1979). However, according to E. Condurachi (1980:119), these latter historians have two precious advantages—namely, the passing of time which allows them the diachronic and synchronic verification of effects flowing from the contacts of civilizations, and abundant and varied data. In reality, the passage of time has one major disadvantage: it hinders the study of the processes of contacts and cultural exchanges as they evolved, which is a grave disadvantage in the search for their causes. As for the data, they are more often abundant rather than varied, and only touch on some very limited spheres. Thus, the few references to the Phoenicians in Greek literary works contemporaneous with the Persian period and the representation of Phoenicians in certain works of Greek art indicate that the exchanges between Phoenician and Greek civilizations in this period did not happen in one direction only, but the gaps in our information do not allow exact measurement of the contribution of Phoenician civilization to Greek civilization (Elayi 1988: 106-107, 147 nn. 4-5). The only thing we can consider with any great probability is that

Greek culture was essentially the donor culture and the Phoenician culture the taker culture; we may note in passing the route followed in the case of traditional Hellenism, which posed this problem in terms of a superior culture and a 'barbarian' culture.

Recourse to comparisons with interpenetrations in modern societies can be useful, on condition that they remain extremely prudent, since there does not yet exist a comparative study between the data of history and the data of sociology, making it possible to determine if there are particular conditions that always lead to the same phenomena, no matter what period might be under consideration. The principal difficulty comes from the impossibility of making a complete study of the exchanges resulting from contacts of all kinds between the Phoenicians and the Greeks, such as modern sociologists can do for a living society. But we can draw the greatest profit from studies on the interpenetrations of modern societies for the understanding of the types of approaches envisaged, the sociological tools utilized and the models established, concerning which we know that they are operative since they have been tested on living societies. It is advisable to choose, between the different schools and methods, those which can be adapted to our specific field and have already produced solid results, while keeping in mind that we are using an incipient science. It is not necessary to try to resolve everything at once, but to set out the problems one after another, in clear terms, to try to find the first elements of a solution and indicate the directions likely to lead to progress in the research. In the example we have chosen, after some examination, the sketch proposed for the study of the interpenetration of civilizations by North American specialists in cultural anthropology seems to us to be the most useful, on condition that it is supplemented by the comments of R. Bastide (Redfield *et al.* 1936; Bastide 1968).

However, it is not a matter of artificially overlaying the ancient data with concepts tested in modern societies. Thus, this sketch does not take into account the gaps that exist in our knowledge of ancient societies; the points which remain out of reach will be disregarded, in order to develop to a greater extent others which seem to characterize an ancient society better, and a different progression will be followed, one better adapted to this specific problem. On the other hand, we will reject the emphasis on culture as practised in cultural anthropology, which postulates the separation of cultural and social matters in accordance with an underlying idealist philosophy that gives pre-eminence to the spiritual elements. It is not a matter for all that of dissolving the social in the cultural, since social matters can change without the culture being modified owing to the power of tradition, and culture can pass from one society to another as if it had an independent existence (Bastide 1968: 318).

It is absolutely necessary to redefine the concepts, the terminology and the criteria according to the problematic being studied, and first of all to find a term that can take into account all the phenomena inherent in the meeting of different cultures. We habitually use the term 'influence', which is handy and covers one of the concepts used the most, under different titles (de Montmollin 1976: 2-8): it is in fact a pseudo-concept, since to ascertain an 'influence' amounts to recording evidence. When we speak of 'influence', we speak of borrowing and adaptation of that which is borrowed, in whatever field it may be; but taking note of a borrowing gives us information neither on its reason, nor on the reason for the transformations that they made it undergo, nor on its success or its failure, nor on the reasons for that success or that failure. Only the study of the social milieu where the borrowing was made could permit a response to these questions, or at least to some of them.

Modern sociology uses the term 'acculturation', as defined by North American cultural anthropology, which has conceptualized it: 'Acculturation comprehends those phenomena which result when groups of individuals having different cultures come into continuous first-hand contact, with subsequent changes in the original cultural patterns of either or both groups' (Redfield *et al.* 1936; Herskovits 1952: 225-29; Bastide 1968: 315). This definition does not take into account the fact that direct contact of groups is not essential, and that acculturation can happen as a result of simple processes of diffusion; continuous contact is not necessary either. The concept of acculturation, which implies a cultural mutation, is not the only way to take into account all the phenomena inherent in the encounters between different cultures; these contacts are liable to produce several kinds of phenomena, which can lead or not to acculturation. Consequently, to designate all these phenomena at any stage of their evolution and without prejudging the meaning of cultural transmission, we prefer a more adequate general term, such as 'intercommunication' or 'interculturality'. In fact, the term 'interculturality' is able 'to encompass a much more extensive and varied range of contacts, exchanges, transformations, new syntheses, valid for any part of the world, for any period of history' (Condurachi 1980: 117).

The choice of data is not done in the same way as in the sociology of modern societies, where the sociologists choose those to be studied and limit as they please the extent of the field of research. The nature of the data at our disposal is the random result of the documents that have come down to us: they rarely make it possible for us to know anything other than some limited cultural borrowings, and spheres that are not at all or little documented are not accessible to us. On the other hand, if the ancient documentation provides starting points for access to intercultural phenomena, it provides no information on the causes or on the evolution of these phenomena. However, being limited to the study of isolated and limited data does not prevent us from trying to grasp better, so far as possible, up to what point it is possible to speak of Hellenization. The phenomena of interculturality ought to be studied as a total social phenomena, namely, under all its aspects, and in dialectical relation with the group to which it is attached (Bastide 1968: 321).

As modern sociologists have clearly understood, it is absolutely necessary to establish first of all a typology of contacts: the five specific criteria proposed take into account at the same time both the nature of the documentation and the problematic chosen. The first and most general one determines the situation in which the contacts were produced, whether it was a situation of spontaneous cultural exchanges, or a situation of controlled acculturation, when an external culture is imposed by various means. It seems to us useless to distinguish, as R. Bastide does (Bastide 1968: 325), two types of controlled acculturation, namely, planned or not. The Phoenician cities never found themselves in the Persian period in a situation of controlled acculturation; in the Greek cities, where the Phoenician community was a minority, the problem was different, but as for the Greeks settled on the coasts of the Eastern Mediterranean, what we know of their installations there does not entitle us to say that the Phoenicians could find themselves in this minority situation. The second criterion is used to measure the homogeneity or heterogeneity of civilizations in contact with others; the presence of numerous foreigners in the Phoenician and Greek cities meant that the milieus concerned were always heterogeneous milieus. The third criterion, established against the background of the specific political situation in the Persian period, defines the contacts between hostile societies (the Median wars, which brought the Phoenicians, allies of the Persians,

into conflict with the Greeks), between societies in commercial competition (Phoenician and Greek trade, in the Eastern Mediterranean in
particular), or between friendly societies (Sidon and Athens, or Sidon
and Salamis of Cyprus, in the fourth century).

The fourth criterion, a fundamental one, is concerned with the relations between entire populations and isolated groups—such as, groups
of Greek merchants or residents in contact with Phoenician merchants
or residents (in North Phoenicia), or with the Phoenician populations
as a whole (Sidon); it seems to us useless to be more specific, as
R. Redfield and J. Herskovits are, on the size of the groups in contact
(Redfield *et al.* 1936; Herskovits 1952: 232). We must be careful not
to forget the indirect contacts between two groups through a third
group, or several other intermediate groups (or through the support
of written work); thus, when the Phoenicians themselves sold Attic
ceramics, they could have been in contact with Greek merchants, or
with intermediaries such as the Phoenicians of Cyprus; we speak in
this case of diffusion. The fifth criterion, likewise fundamental, defines
the place of contacts: coastal or hinterland, urban or rural milieus,
Central Phoenicia, North or South Phoenicia, zones with a dominant
Phoenician presence, zones with a dominant Greek presence or mixed
zones; R. Bastide prefers to reduce this criterion to that of closed or
open societies (Bastide 1968: 326). All things considered, it is relatively
easy to draw up a typology of cultural contacts, even if future research
is bound to refine the analysis.

It is far more difficult to establish a typology of the results of these
contacts, in other words, of the different processes produced by the
phenomena of interculturality. Modern sociologists have proposed a
whole series of designations;[3] historians of religion have especially
spoken, in regard to borrowings of religious cultural features, of syncretism under various forms. P. Lévêque distinguishes four such forms:
borrowing, juxtaposition*, reorientation* and crasis* or amalga-
mation (Lévêque 1973). Perhaps it is for lack of sufficient precise data
that these categories are only in part applicable to the phenomena of
interculturality observable in Phoenician religious beliefs in the Persian
period.

When the Phoenicians made use of Greek objects in their rituals, we
must decide whether their usage implied a radical modification of these

3.   Reinterpretation, reorientation, readaptation, indigenization, syncretization,
reconstellation and synthesis.

rituals, or whether they were treated simply as interchangeable instruments. What was the significance of the depositing in the tombs, side by side with local objects, of Greek vases on some of which Greek myths were depicted, while others were funerary vases, such as white ground lekythoi? The Phoenicians would very likely be acting in this case like the African Bantu, who borrowed Catholic ornaments for simple aesthetic reasons; in this juxtaposition of heterogeneous cult objects, the two religious spheres did not interfere with one another. When the Phoenicians offered to their gods terracotta statuettes representing Greek deities, they went beyond the simple juxtaposition of cult objects. Once again, the study of an analogous phenomenon among the African Bantu can provide some elements of an explanation: when the latter replaced their *pegis* (local gods) with statues from the Catholic religion, the two religious spaces had become just one. That does not mean that morphological syncretism always automatically finds expression in the identification of the *pegis* and the saints, but it furthers it, through the confusion that it introduces into the feelings and imagination of the faithful (Bastide 1960: 380-81; Elayi 1988: 137, 153 and n. 95). The Phoenician documentation seems to indicate that the Greek gods represented by the statuettes resembled the Phoenician deities to whom they offered them, or at the very least were not incompatible with them; establishing an equivalence between these deities was at least possible starting from the morphological analogy. Occasionally, too, the portrayal of the Greek gods included attributes unknown in the corresponding Phoenician gods: for example, the lion skin of Herakles had a Greek mythical meaning, but nothing like it is known in the case of the representation of Melqart. Perhaps the Phoenicians did not take into account attributes without any equivalents in theirs; but, since these cases are frequent, it seems on the whole that there had been a blending of Phoenician and Greek religious elements through forms that were specifically Greek or Greek derivatives. This research must be pursued, but it still only reveals the starting points of religious borrowings, and not the social and mental backgrounds.

The principal difficulty in fact lies in determining the causes that set in motion the processes resulting from cultural contacts. Even modern sociologists have a great deal of trouble in discovering certain factors, which, can, however be at work to intensify the processes or to neutralize them; they essentially distinguish two causalities that enter into a dialectical relationship: external causality and internal causality (Bastide 1968: 327 and n. 2). The processes of interculturality must be considered as total social phenomena, in the sense that the change cannot be limited, with the internal causality being stimulated by the external causality; it does not remain localized, but has repercussions on everything, since a new equilibrium must be established between the unbalanced parts. Recently it has been shown that the kings of Sidon such as 'Abd'ashtart (Straton in Greek) were Hellenized kings like the kings of Salamis in Cyprus, that in all likelihood they spoke Greek, and welcomed the artists, the political refugees and all the Greeks wanting to come to their court. These kings would have been banished or marginal individuals if they had not been at the centre of the social structure, which means that instead of being rejected by the traditional society in which they lived and which had not undergone the same processes of transformation, they had helped it transform itself. The social level of rich Sidonians was also affected by this cultural change, which, in turn, furthered the diffusion of Hellenism in the whole Phoenician population, particularly through the monetization of trade, through the importation of Greek luxury products and through the creation of a new way of life. We cannot say to what extent the traditional organization of Phoenician society was weakened at the end of the Persian period, but we note that it certainly was, in different degrees according to the places and the social levels considered, since the Phoenician city was a total society where chain reactions were inevitable.

Even if we are still at the experimental stage in the field of historical sociology, it seems clear to us that it is the historians themselves who must establish, with the help of modern sociology, the foundations of this new discipline which represents an indispensable complement to history, and which would be especially valuable in the study of the multicultural milieus of Achaemenid Transeuphratene.

Chapter 11

CENTRAL AUTHORITY AND LOCAL AUTHORITIES:
A MATTER OF PERSPECTIVE

Redefined through new orientations and tested methods in the study of
economic and social phenomena, political history can tackle the fundamental question of governments in Transeuphratene in the Persian
period, such as nature, forms, interactions, break-ups, continuities, and
stabilities. Transeuphratene comprised at that time an extraordinary
mosaic of governments of various kinds, which the historian must give
prominence to and analyse in order to try to discover their underlying
workings, the role of socio-economic interests, and, vice versa, their
impact on economic and social structures. One of the most characteristic phenomena in this region was the overlapping of powers that
sometimes strangely resembled a system of nesting boxes. Let us mention as an example the city of Ashkelon, which was apparently the seat
of a monarchy whose nature still needs to be defined; this city depended
politically on the city-state of Tyre in a dependence that also still needs
to be defined; as for Tyre, it was under Persian rule, through the intermediary of a provincial satrapic government (Elayi 1989: 94, 104 and
n. 112). The same kind of overlapping of powers seems to be found in
Cilicia, a province neighbouring on Transeuphratene, where the Greek
coastal cities such as Celenderis depended on the kingdom of Cilicia,
itself under Persian rule through the intermediary of the satrap of
Cilicia.[1] Before being able to analyse such complicated political situations, we must begin by trying to study separately each political component, while keeping in mind that there is much left to do in this field.
The major drawback in this study is that it is divided up among historians with different specialties who sometimes exaggerate the importance of the government that they know and analyse.

---

1.   With regard to Cilicia in the Persian period, see Lemaire and Lozachmeur 1990.

Thus, an Iranist historian sometimes has a tendency to overvalue the presence of the Persian power in Transeuphratene, by arguing especially from the notion of an imperial–tributary State with a strong territorial hold and from generalizations about models of occupation elaborated for the better documented provinces. Conversely, the regionalist historian sometimes has the tendency to underestimate the indications of Persian territorial ascendency, which have to be interpreted in the light of informative parallels when they are not very explicit. It would be just as unscientific and ideologically suspect to expunge the indications of the ascendency of the central Achaemenid government as to see it everywhere without sufficient evidence, as Flaubert said about the writer in his work: 'present everywhere, and visible nowhere' (letter to Louise Colet, 9 December 1852). The 'centralist' thesis does not seem any more applicable to the situation of Achaemenid Transeuphratene than the 'autonomist' thesis, in any case not a priori. The documentation must first of all be collected together and then analysed in as objective a way as possible. It is here that inter- or multidisciplinarity must play a part: the scholars on Iran must converse with the regionalist historians, or at least each of them should become familiar with the discipline of the other in order to understand their own better. Finally, while the idea of centralization is kept in mind, it is necessary to be aware of the great diversity in local governments and in their socio-economic and cultural aspects, not only from one province to another, but within the same province, and be aware as well of the various expressions of Achaemenid power that cannot be isolated from their social and spacial-temporal dimensions. Thus, the privileged relations established between Xerxes I and the king of Sidon during the second Median War had nothing to do with the conflictual relations between the inhabitants of Sidon and the resident Persian military preparing an expedition against Egypt about 350.

Let us take the example of Persian domination over the Phoenician city of Sidon, which has left very few traces and which seems completely different from the type of domination known in the well documented satrapies of Asia Minor or Babylon (Elayi 1989: 135-59 [with bibliography]). The main preoccupation of the Achaemenid kings was the collection of tribute and the various taxes. Submission to tribute was not a new situation for Sidon, which had already been a tributary of the Assyrian and Neo-Babylonian kings, but the tribute seems not to have been really regular before the end of the reign of Darius I. Its regular collection during such a long period (more than a century and a half) was a new phenomenon which had to affect in a particular way the economic life of Sidon. Although the Persian tribute was estimated by Herodotus in money value, it must have been, at least in part, paid in kind as it was during the Assyro-Babylonian domination, for example, in cedar wood to be used in the construction of the palace of Darius at Susa. It also had to provide the wood needed for the construction of the powerful fleets used in the service of the Persians, which would represent a considerable supply in a militaristic state like the Achaemenid Empire.

Besides the tribute, Sidon was subject to other taxes, less well known, such as an assessment on the working classes, a conscription rendered particularly heavy because of the naval power that was the backbone of the Persian Empire, and so on. Other burdens weighed heavily on the city: its artisans could be conscripted by the Persians to work on the spot, in the arsenals for example, or in the great cities of the Empire for the construction or restoration of palaces. The Sidonians also had, in one way or another, to participate in the support of Persian residents and troops stationed there or passing through; perhaps it was as a sign of protest against this tax that about 350 BCE they burned the forage intended for the horses of the Persian garrison. The Persians also exploited the enormous maritime capacities of the Sidonians by using them as commercial intermediaries: they were specialists particularly in commercial transportation and could supply the Achaemenid court with products which were not available in the Near East, even the most distant and most rare ones.

Finally, the Persian kings had reserved for themselves some land in the territory of Sidon; it was the famous royal estate which the Greeks called *paradeisos*, whose name in Phoenician we do not know. The one in Sidon, of which Diodorus speaks, must have been situated on the western slope of the coastal range, which towered above the city, since it included in all likelihood cedar forests as in the royal Assyrian reserve. This *paradeisos* probably had as its purpose to preserve the most beautiful trees for exploitation by the Persian kings, to serve as a pretext for levying taxes, and must have formed a hunting preserve, an aristocratic diversion which had an important place in the ideology of these sovereigns. There is no reason to think, however, that they directly intervened in the administration of the civic lands of the city of Sidon or in their means of production. A still unedited inscription of King Bod'ashtart of Sidon, apparently about the construction or the repairing of the valuable Saltana canal, which probably supplied several sectors of Sidon, shows that the control of the supply of water belonged to the city; in Mesopotamia, on the contrary, royal control in the area of irrigation and river transport was a permanent feature, but the Persian sovereigns, unlike their Neo-Babylonian predecessors who conceded the upkeep and use to some large bodies such as sanctuaries, seem to have had a state monopoly over this, which assured them in this way a control over agricultural activity and a source of revenues. Nevertheless, while the control of water remained with the Sidonians, the presence of a Persian residence in the very territory of the city of Sidon implies that the Persian nationals would have possessed close by, to provide for their support, some royal lands, probably administered according to the Achaemenid modes of production. We do not know whether these lands formed part of the *paradeisos* mentioned by Diodorus, or if they were separate.

The regularity of the collection of tribute and of various taxes was dependent on a good territorial organization. We know that Persian troops were stationed at Sidon, at least in the middle of the fourth century. The Persians seem to have been interested in the territory of that city, either as a strategic location to control the coastal route, or as a base of operations on the road to the Greek cities or Egypt, which always presented problems for them; there was, therefore, in the city, toward the middle of the fourth century, an increased Persian concentration. But even in that period, the Persian garrison at Sidon must have been quite limited in number and in means, since it witnessed, powerless to intervene, the Sidonians making preparations for war. It does not seem to have been a real Persian military colony at Sidon, but a certain number of troops in the service of Persian residents, which had to symbolize the military power of the occupiers. In case of need, when the unity of the Empire was threatened, the Achaemenid kings did not hesitate to strike very hard, as the violent repression by Artaxerxes of the revolt at Sidon about 350 BCE shows.

The Persian domination at Sidon was accompanied too by good politico-administrative supervision, inspired by the examples of previous Empires. The Persians had, like their Assyrian and Neo-Babylonian predecessors, a double control system, namely, periodic and permanent. The periodic inspections of the satrapies were carried out by a high-ranking civil servant, specially appointed by the Great King and travelling with a Persian military escort. This periodic control, which was already prejudicial to the autonomy of Sidon, was reinforced through control of the principal routes and a permanent control within the city, the exact nature of which is difficult to evaluate for the moment. The ideological representation of the Achaemenid power at Sidon escapes us as well.

As a general rule, the Achaemenid kings used peaceful means for their domination. Moreover, among the different states under Persian rule, Sidon fitted into the most privileged category, namely, that of the autonomous tributary states, which were able to retain their political officials, under certain conditions. In view of the power of the city and the interest it had for the Persians, the king of Sidon received preferential treatment; under the reign of Xerxes I, he enjoyed the principal consultative role in the Persian war council and was treated as a privileged collaborator. Sidon enjoyed a relative autonomy, on condition that it paid a regular tribute, a symbol of subjection, placed its forces at the disposition of the Persian kings and did not act to the detriment of the interests of the Empire or compromise the peace. Even when the Persians used Sidon's navy, their control remained indirect, since it was built and armed by the Sidonians, provided with a Sidonian crew and commanded by the king of Sidon. In order to maintain the loyalty of the kings of Sidon, the Persians continued to practise the already ancient system of the gift of land and favoured the persophile dynasties.

On the whole, politico-administrative surveillance was present at Sidon, although not as rigorous as elsewhere. Even if there had temporarily been a satrapic seat in the city, the present documentation shows that there had been neither a military colony, nor a colony of Persian population there, as was the case in other regions of Asia Minor, but only a minority of Persian residents, probably on good terms with the king of Sidon and those near him favoured by the Great King, but having limited contacts with the local populations in order to preserve their own ethno-cultural and political specificity. It is obvious that the Sidonian example cannot be generalized: we will have to study as well how the Persian domination was exercised in other states of Transeuphratene.

Unlike Achaemenid Babylon, which had suffered a gradual erosion of traditional political and administrative powers, especially beginning with the reign of Xerxes I (Joannes 1990), Achaemenid Transeuphratene seems by and large to have better preserved its local powers. These powers, varied in nature, were not all treated in the same way by the central Persian power and the status of each of them had undergone modifications as well in the course of two centuries of Persian domination. We will take up again the example of the city-state of Sidon, while comparing it with the very much differentiated local powers of Judaea and the kingdom of Arabia.

The example of Judaea, still difficult to evaluate, illustrates the case of an apparently weakened local power. The status of this territory is very poorly known; while we can approximately piece together its extent (Lemaire 1990), we do not know what politico-administrative meaning we should give to *Yehud medineta'*, 'province of Judaea'; the hypothesis of A. Alt, according to which Judaea was attached to Samaria up to the time of the mission of Nehemiah (Alt 1934), seems to be invalidated today by new documents (Williamson 1988; Lemaire 1990) to which we will refer later; even then, we would have to be more certain about the meaning of some keywords. We possess on the other hand some pieces of information on local seals and coins in regard to the government functions carried out by some of the persons mentioned in the Old Testament, without being able, however, to measure their probable evolution during the two centuries of Achaemenid domination.

The power exercised by Nehemiah is up to the present the best known. He was obviously given responsibility by the king of the Persians for a specific mission of a politico-administrative nature, limited in its content and duration, namely, to rebuild and repopulate Jerusalem as a fortified city and capital of a 'province', so as to restore social order and ensure the proper functioning of the cult. Nehemiah accomplished this mission as a representative of the Persian government, more precisely of King Artaxerxes I. According to the presentation of the book which bears his name, Nehemiah carried out, in the same reign, a second mission of a more socio-religious character than the first, namely, sorting out the administration of the Temple, reasserting the value of the Sabbath and of marriages between Judaeans (Neh. 13.6-31). We clearly have the impression, however, that through his person it was the Jewish community of the Babylonian Diaspora who, from a distance, imposed its views on the reorganization of Judaea. The kind of power exercised by Nehemiah to impose new socio-religious rules in the setting of his second mission contains nothing surprising; it seems to have been expressed through a mutually agreed-upon contract, sealed (that is to say, signed) by the religious personnel, priests and levites* on the one hand, and by the heads of families, representing the people on the other; by counter-signing the contract himself, Nehemiah was its guarantor, which means that the central power, through its duly mandated agent, intended to keep control of the internal affairs of Judaea (Neh. 10). We get further insight into the weakness of local powers in the account concerning Ezra. No matter what relationship we can establish between the missions of Nehemiah and Ezra, the latter, being a priest and a specialist in the 'Law of the God of the heavens', had been sent by the Great King on a juridical inspection; through his mission, that Law was going to be enforced in Judaea as a 'Law of the King' (Ezra 7.11-26). On the other hand, the account of the book of Ezra, which centres his mission on the '*Torah* of Moses', its reading and exegesis, seems to place the religious reorganization of Judaea within the compass of the Babylonian Diaspora.

Nehemiah had the title *peḥah* in Hebrew and Aramaic according to the biblical texts, a title also found in the inscriptions, seals and stamps coming from various finds (Aharoni 1962: 7-10, 32-34; 1964: 21-22, 44-45; Avigad 1976: 6). This term is usually translated by 'governor', but it seems to have had a broader meaning, designating royal officials of varying ranks, assigned political and administrative missions in different areas of the Empire. After Nehemiah, at least two other persons would have had this title in Judaea: one with a Persian name, Bagohi, according to a letter from Elephantine in Upper Egypt; the other with a Yahwist* name, Hezekiah, according to coins dated to the end of the Persian period. Before Nehemiah and at the beginning of the Persian period, two individuals mentioned in the Bible, Sheshbazzar and Zerubbabel, seem to have been invested, under the same title, with limited missions like those of Nehemiah, to carry out in particular the reconstruction of the Temple and the restoration of the sacrificial cult at Jerusalem. Despite their Babylonian names, they seem to have been authentic representatives of the deported Davidic line and therefore chosen from within the community in exile as bearers of an ancient local authority and invested with a new authority, delegated by the central government (1 Chron. 3.18-19; Ezra 1.8; 5.14-16; Hag. 1.1). In the case of Zerubbabel at least, this method of handling local governments, extremely common in the context of the Empires of the ancient

Near East, led to Judaean political currents nourished with messianic hopes (Hag. 2.20-23; Zech. 4.6-10) 'playing' the Davidic line against the Persian authority. The seal of Shulamit, who seems to have been the daughter of Zerubbabel, would show that the Persian chancellery, at the beginning of the fifth century, relied again on that local authority in naming Elnathan, her husband, 'governor' in his turn, but showed a certain prudence from then on by moving away from the Davidic line (Avigad 1976: 5-6, 11-13).

Other seals, discovered at Ramat Rahel in an embankment, and therefore difficult to date, give two other names of bearers of the same title; from the paleographic point of view, they can be dated to the fifth century, a little before or a little after the time of Nehemiah (Aharoni 1962; 1964; Lemaire 1990). They apparently were used to seal jars of wine or oil, an operation in which the owners, as official personages, seem to be guarantors for the contents; it is not impossible, then, that in addition to their politico-administrative responsibility, they would have had to supervise the administration of the royal domains meant to provide for the needs of local administration and/or the royal court. As for the collection of taxes, at least of taxes in kind, this seems to have been within the competence of the 'governor' or of members of his family, as would be indicated by a batch of bullae and seals, which, because of the references to the name of the 'province', Yehud, and of the presence of the seal of Shulamit, take on the value of an official archive.

The Persian chancellery would probably choose its officials mandated in this way from the most loyal and easily controllable families, that is to say, preferably in the Diaspora close to the central power. They put up with politico-administrative and socio-religious initiatives like those of Nehemiah, the 'governor', and Ezra, the 'scribe of the Law of the God of heavens' (and of the King), insofar as they reinforced the internal cohesion and the fidelity of Judaea to the central power. We note, however, that there is no indication in Judaea of the establishment of a dynasty of governors, as was the case in Samaria, which did not present the same risk from messianic temptations and therefore from revolts.

Unlike Judaea, the city of Sidon had preserved its autonomous sociopolitical structures and constituted as it were a State within the State (Elayi 1989 [with bibliography]). Contacts between the Great King and the city were generally not direct, but passed through the satraps or governors from the Persian side and through the kings of Sidon from the side of the city. Certainly, royalty constituted a characteristic element of Sidonian political institutions; it was a unique royalty, for life and hereditary. The succession to the king could take place through his son, but also through his brother or through a close relative; when Alexander chose to replace the king of Sidon, not with his son, but with another member of the royal Sidonian family, he was respecting the local procedure known to us in other ways, as he ordinarily did in his conquests. The kings of Sidon sometimes practised consanguineous marriages, probably with the principal aim of strengthening the royal line.

The kings of Sidon had more or less well-known priestly, military, judicial and diplomatic functions. The priestly function is the best known and was probably the most important. Some kings of Sidon, such as 'Eshmun'azor I and Tabnit, were priests of 'Ashtart; the place given to this priestly title in the royal titulature, before the royal title and the patronymic, indicates that it covered a religious function that had priority. That function was also carried out in constructing, rebuilding or decorating religious edifices in the city: thus, King Bod'ashtart ritually buried several copies of commemorative inscriptions in their foundations. Establishing new places for cult was another royal prerogative: thus, King 'Eshmun'azor and his mother installed the divinities Astarte and 'Eshmun in a new cult place. The kings of Sidon could also personally perform priestly acts, as seems to be indicated in a scene represented on the reverse of local coins, which has not always been well understood. They seem, in a word, to have used religion to maintain their political power. In this way, the works they carried out in the temples not only had as a goal to assure themselves divine favour and divine protection; they were also a propaganda tool intended to give to the inhabitants of the city a certain image of the king as pious and magnificent. The justification of royal power through religion is qualified in the Sidonian inscriptions, addressed ostensibly to the gods, but without doubt meant to be read by all; whence the importance of recalling the divine legitimation and protection of the kingship.

The military function of the king of Sidon is not expressed through a particular title, but is well known from the Classical sources and the images on coins: the king commanded his city's fleet. His authority over the fleet was not just theoretical; in time of war, he became a real admiral who prepared naval strategy and participated in battles; there was no separation between the direction of the affairs of the state and the war at sea. The king was also commander of the army, at least at the end of the Persian period. The Sidonians, whose armies had been used by all the successive conquerors, had long military experience and the importance of war in their city is illustrated by the presence of a warship on their coins. However, the military obligation imposed by the occupier, while it favoured military development, did not leave much room for specifically Sidonian undertakings. The military authority of the king of Sidon was not limited solely because of Persian domination, but had its limits as well even within the city itself. The judicial and diplomatic functions of the king are less well known.

There existed at Sidon, in the Persian period, beside the royal power, another power, held by the citizens, which was more or less secure and no doubt in certain cases dominant. Far from having a monopoly on the priestly, military, judicial and diplomatic functions which he exercised, the king shared them all, in varying degrees, with the citizens of his city; the small extent of Sidonian territory must have favoured their participation in the affairs of the city. Beside the power of the king, some members of the royal family, such as the son and the mother of the king, could also have held political power, which was exercised, for example, in the form of a coregency. There existed several representative organisms, more or less limited and selective: perhaps an enlarged assembly, a Council of Elders and a Council of 'The Hundred'. The balance between the royal power and the representative bodies, or between the various bodies was sometimes precarious. In some apparently exceptional cases, a select committee could have intervened in the choosing of a new king, and the decision could then have been simply ratified by the people or the assembly of the people.

The evolution in relations with the Persians played some role in Sidonian politics. Thus, after its revolt in 350 BCE, this city lost its hegemony over the other Phoenician cities; on the other hand, the disappearance of a large number of Sidonians, especially of members of representative bodies, combined with Persian political pressure, seems to have led to the institution of a tyrannical regime with Straton II; the civic community only regained political power with the arrival of Alexander.

Northern Arabia, lastly, was not at all treated by the Persians in the same way as Judaea and Sidon, because it did not form a homogeneous whole, but was made up of tribes and cities (Stern 1982: 250-52; Knauf 1990; Lemaire 1990 [with bibliography]). The tribes were looked upon as clients*, while the cities were more firmly integrated into the Persian administrative system. Even when we know the name of a Persian governor of Arabia, that does not imply that Arabia formed by itself a satrapy; it was probably attached administratively to Transeuphratene. The cities which had been, as far as we know, integrated into the Persian administrative system were four in number: Bosra, the Edomite capital; Elath (Tell el-Kheleifeh) on the Gulf of Aqaba, uninterruptedly inhabited from the sixth to the third century; Dedan, one of whose local governors we know; and Tayma where there has been recovered, among other things, a stele commemorating the erection of a throne by a still quite mysterious personage.

With regard to the tribes, we will mention first of all that of Qedar which controlled, beginning in the second half of the fifth century, the South of Palestine, the Sinai, the South of Transjordan and the Hijaz; the chiefs of this tribe referred to themselves as kings (Knauf 1990). The tribes of Nebaioth, Massa' and Hadad, enemies of Tayma, have not yet been localized; the same is true of the tribes of the Hagaraeans (?), of Naphis, Nodab and Yetur. There seems to have been in the Persian period a kingdom of Lihyan situated more to the north.

It is interesting to note the existence of clients at two levels, in the sense that the chiefs of tribes could have been at the same time clients of a neighbouring powerful chief and, through him, clients of the king of the Persians. Thus, the king of Qedar controlled a good part of the Arab tribes and served as an intermediary for them with the central Persian power. The city of Tayma, too, seems to have controlled the neighbouring tribes.

It seems that the whole region had undergone important changes relating to political control between the end of the fifth century and the beginning of the Hellenistic period; the Nabataeans perhaps made their appearance, more or less taking the place of the Qedarites, and a new province was created, namely, Idumaea. The most probable hypothesis would amount to linking these changes to a reorganization of the south of Palestine by the Persians in the fourth century. The successive revolts that shook the region can explain why the Persian authorities had felt the need for an administrative and military reorganization, perhaps in the energetic reign of Artaxerxes III. This reorganization seems to have been carried out at the expense of the Arab kingdom of Qedar, which would have taken part in the Cypriot revolt of Evagoras, who had the support of the Pharao Achoris. The extent of the new province of Idumaea at the time of its creation is difficult to determine precisely; probably it included all the territory to the south of the province of Judaea, from Lachish, Maresha, Hebron and Ein-Gedi, as far as the Negev; its capital has not yet been identified. It was probably also at this time that the city of Gaza was placed under the orders of a Persian military governor.

These three examples give some idea of the richness, variety and complexity of the political history of Transeuphratene and of the immensity of the task to be carried out in this field alone; we have begun to deal with this task in our first International Colloquium, but there are countless others that must still be tackled.

CONCLUSION

As we come to the end of this study, the attentive reader will no doubt have noticed that we have said nothing about Cyprus, although it was presented in our Introduction as forming an integral part of Achaemenid Transeuphratene, even if the details of that attachment still remain to be defined. It especially should not be considered forgetfulness on our part, since to tell the truth Cyprus is at the centre of our preoccupations, and it is for this reason that we have chosen to end our book there.

Cyprus is situated, as everyone knows, on the frontier of two worlds, Near Eastern and Hellenic, and was the privileged place of their impact and of their interpenetration in the Persian period, representing for historical sociology the ideal laboratory for the analysis of the processes of interculturality. Furthermore, its documentary potential is extensive, compared to that of many of the sectors of Syria-Palestine, as much in public collections as in the results of excavations already published and in archaeological reserves still unexploited; although not as large, the textual material and the epigraphic collections are far from negligible and still have much to contribute. Knowledge of the Persian period in Cyprus constitutes in our opinion an obligatory stage in any study relative to Achaemenid Transeuphratene, and the reverse seems to us equally true.

Without going into details, which would be too long and sometimes delicate, it seems to us necessary to take a bearing briefly on the main problems that come up. In a general way, almost everything we have just written in regard to Syria-Palestine applies equally to Cypriot studies, sometimes still more emphatically even, and it is therefore useless to return to it; we will be content to stress some more specifically Cypriot problems. We evidently could not deny the interest in the results obtained up until now in Cypriot studies, which are sometimes very important, but when we put them in perspective, taking into account the rich documentary potential, the facilities for research provided and the means brought into play, all are not perhaps always at the level at which they should be, especially in regard to the Persian period, which interests us here. Thus, to read some works, we get the impression that the main objectives have not changed that much since the time of Marquis L.P. Di Cesnola and that they always give priority to the collection of the most prestigious or original objects, meant to supply the collections open to the public today, which is highly commendable in itself, but of slight benefit for advancing the social sciences. The leitmotiv of cultural contact recurs in most of the works, but it seems that they speak of it most often because the subject is fashionable and because it is imperative in a multicultural Cypriot milieu, and not so much because of an interest in really renewing the approaches and perspectives.

Such a situation is explained in part by the fact that Cypriot studies are today mainly in the hands of Hellenists, at least with regard to the Persian period. If it is entirely legitimate that they actively contribute to research in this field, it is hard to understand why there are so few Orientalists to take over Cypriot Phoenician studies, which are of no less scientific interest than Cypriot Greek studies, insofar as they can be separated. In fact, it amounts to one of the most closed fields of research on Antiquity: starting from Cyprus, we can have access to studies on the Near East, which is entirely normal (King Evagoras I of Salamis had already landed there in the fourth century BCE!), but the reverse journey is almost impossible, in particular where documentary access is involved. This attitude of isolation puts a serious strain on research since studies on Achaemenid Transeuphratene, including Cyprus, cannot really progress without close collaboration between the researchers on both shores. A mere hundred kilometres separate Ras Ibn-Hani, to the north of Lattakia in Syria, from Cape Andreas on the eastern end of the island of Cyprus; we would hope that modern researchers could cross in both directions that expanse of sea which in the time of the Persians joined two parts of one and the same political, cultural and economic whole.

This, in fact, is the spirit of this book—namely, to reject barriers and frontiers, of whatever sort they may be, that might get in the way of scientific advances. This book is certainly not, for all that, an indictment (by what right would it be) of the methodological errors and the 'imperialisms' of disciplines or of fields, but the development of an open point of view, which has no other objective than a scientific one.

We have not really tried either to draw up a balance sheet, since the results so far achieved are still too incomplete, although very promising, or a series of solutions to the problems that have come up, which would be quite premature. This book is presented as an ongoing inquiry, on all levels and from the most diverse angles, in a new field of research. We would be satisfied if this approach could make it possible to clarify the starting points for reflection, to eliminate non-issues and to define the real problems. The questions set out have sometimes prompted other questions, sometimes led to answers, and sometimes showed the way to formulating hypotheses and to proposing the general orientation of research. One could detect at times in our remarks, judgments and proposals, a certain level of provocation; the only objective intended was to elicit reactions, whatever they might be, but preferable in any case to apathy or to the conformism in which research sometimes takes refuge. We have tried to show in a non-complacent way that studies on Transeuphratene in the Persian period, as promising as they could be, are still today a vast construction site where almost everything has still to be built and a laboratory where many methods remain to be tested. Real and well-understood multidisciplinary and interdisciplinary methods seem to us the only way to make progress in this new research field, and we invite all interested researchers whose only ulterior motive is scientific, to join our group in a spirit of openness, of dialogue and of dynamism, which we have already been trying to practise for some years, through consideration of the acquired knowledge of the past, the illumination of the new approaches of the social sciences, the assistance of the apparatus of modern technology, and respect for the opinions of others.

This new look at Achaemenid Transeuphratene has allowed us to give prominence to a new field of research that we have tried to mark out by appraising the distance already covered, the research in progress, and the future prospects, in order to specify as much their contributions as their already discernible limits. Probably several of our assessments will be rapidly surpassed by works of practical experimentation as well as by those of theoretical reflection applied to the field which has occupied us in this book. This is a development that we desire, since it will prove that research on Achaemenid Transeuphratene is really alive, productive and better and better adapted and equipped to document and study this subject. Paradoxically perhaps, we think that the more quickly our 'new look' will lose its novelty, the more our enterprise will be shown to be useful; since, as Lucien Febvre said, the usefulness of a book 'is all the greater as it is more quickly replaced by the many research projects that it suggested or provoked' (Febvre 1954a: 145-47).

GLOSSARY

Absolute Chronology The dating of documents according to our present method of computing, by referring to documents that can be precisely dated; in archaeology, the objects found in situ* and dateable in this way make it possible to move from the relative chronology* of strata to their being fitted into the absolute chronology.

Acculturation According to the definition of R. Redfield, all the phenomena which result from the contact between two groups of individuals of different cultures, with the changes which take place in the original cultural models of one or two of the groups.

Achaemenid Refers to a Persian dynasty named after its ancestor, Achemenes, and founded by Cyrus II the Great about 550 BCE; it extended its Empire over the whole Near East and was brought to an end by Alexander the Great in 330.

Akkadian An East Semitic language current in Mesopotamia from the middle of the third millennium BCE. Adopting Sumerian cuneiform writing, old Akkadian diversified into Babylonian in the south and Assyrian in the north and lasted until the beginning of our era.

Anthropoid Sarcophagus A sarcophagus whose lid and/or coffin recalled the human form. The production of anthropoid sarcophagi in stone, mainly Sidonian, probably lasted from the second quarter of the fifth century up to the end of the Persian period.

Architectural Sarcophagus A sarcophagus whose coffin and lid, adorned with sculptured motifs, had an architectural form bringing to mind a house. The examples known from Transeuphratene are those of Amathus, of Golgoi and especially of Sidon.

Arkhē A district of government.

Balk A strip of earth left in place along the grid lines of an archaeological site (see grid system*), limiting the size of the excavated square; the four banks which border the square serve as material witnesses and samples of the stratigraphy*.

Bronze Age A traditional chronological phase in archaeology, subdivided as follows for the Near East: Ancient Bronze (3300–2000 BCE), Middle Bronze (2000–1550 BCE) and Late Bronze (1550–1200 BCE).

Bulla A small lump of clay applied to the cord tieing up a papyrus; the clay bears the imprint of a seal on its external side and can retain the traces of papyrus fibres and of the cord on its inner side.

Canon 'Canon of the Old Testament' refers to the list and the definitive form of books and collections of books which, at the beginning of the Christian era, received in the Jewish community a sacred authority as a result of their divine inspiration.

Client In studies on Arabia of the Persian period, this term refers to any person who is placed under the patronage of a person of higher rank or of one who is more powerful; a very powerful chief of a tribe had as clients other chiefs of tribes and was himself a client of the king of the Persians.

Comparator of Dies A device used to compare coins two by two (originals or photos) to help in the identification of identical dies.

Concordance An alphabetical table of words found in the Bible, referring to all the passages where they are cited; concordances have been drawn up for other collections of texts and inscriptions as well.

Countermark A mark struck on a coin after its manufacture, either to modify the value of the coin, or to give it currency in another context than that of its issue.

Crasis According to the typology of syncretisms established by P. Lévêque, a blending of two religions causing a different unity to be formed from its two constituent parts.

Die A block of iron or bronze, bearing on its concave side the engraving for one of the two sides of a coin and used in its manufacture by being struck with a hard blow by a hammer.

Die-Linking Establishing a temporal sequence between two series of coins based on the similarities of dies* for the obverse and reverse sides.

Documentary Hypothesis An attempt to reconstruct historically the redaction of the Pentateuch through the identification of sources which constitute continuous and homogeneous documents.

Econometrics An analytical method by which quantitative data are subjected to a mathematical processing, in order to extract connections of a mathematical nature among certain economic phenomena, valid in a predetermined institutional setting.

Emporium A Greek trading post or commercial establishment set up outside the Greek world.

Expert System A group of software programs making use of precise knowledge relative to the area of expertise (for example, that of archaeology), to provide a result comparable to that from a human expert.

Flan A metal disk on which the imprint of two dies* will be stamped on the obverse and reverse to form a coin.

Geochemical Prospecting In archaeology, prospecting in an occupation level in order to bring to light the structuring of the site studied, through the analysis of certain chemical elements coming from heavy human or animal occupation (phosphates, phosphorus, organic material, etc.).

Geophysical Prospecting Prospecting that uses the methods and techniques of analysis of the physical properties of soils (electric, magnetic, electromagnetic resistance), which by contrast helps in the detection of buried archaeological structures (buildings, tombs, graves, etc.).

Grid System In archaeology, a grid pattern for the whole area of a site with tightly stretched strings (with stakes) to prepare for controlled surveying or excavation.

Hypogeum Tomb A subterranean burial place, generally cut into rock and made up of one or several rooms, with an access passage or shaft, which rich families had built to deposit there the sarcophagi of their deceased.

In situ A Latin expression which, in archaeology, describes objects found 'in place', that is to say, in the context of their ordinary usage and in a sure stratigraphic situation.

Integrating Nomadism A form of socio-political relations developed in the societies of the semi-arid regions of the Near East, where the federated nomadic tribes would take over political control of the area, by integrating the sedentary populations that had their own local governments.

Interculturality According to the definition of E. Condurachi, this term makes it possible 'to encompass a much larger range and variety (than the term "acculturation") of contacts, exchanges, transformations, new syntheses, valid for any part of the world, for any period of history'.

Iron Age A traditional chronological phase in archaeology, subdivided in the following way for Palestine: Iron I (1200–900 BCE), Iron II (900–540 BCE); and for Syria: Iron I (1200–900 BCE), Iron II (900–700 BCE), Iron III (700–540 or 700–333 BCE).

Isolate In anthropology, this word refers to a cultural survival.

Juxtaposition According to the typology of syncretisms established by P. Lévêque, juxtaposition consists of the introduction of a foreign divinity into a pantheon, by giving it a local appearance.

Karum An Akkadian term which referred to 'the quay', and, by extension, to the local market and its public buildings.

Levite A member of the priestly personnel of subordinate rank, serving in the sacred Temple of Jerusalem (postexilic period), whose origin is obscured because of contradictory biblical evidence.

Local Scale The scale in a study of a geographical area having an extent of one hundred square metres.

Macroprovenance (Study of) In ceramography, research on the more or less distant provenance of vases imported into a given region, making it possible to reconstitute the commercial, and often maritime, routes.

Massora A tradition on the written form, pronunciation and rules for reading of the Old Testament, fixed principally in the rabbinical schools of Tiberias and Babylon, between the sixth and twelfth centuries of our era.

Microprovenance (Study of) In ceramography, research on the provenance of bowls in the interior of a region or micro-region, aimed at reconstituting the economic circuits involved (production, distribution, utilization).

Nomos See *arkhē*.

Northwest Semitic Inscription In the Persian period, any inscription in Phoenician, Punic, Aramaic, Hebrew or Edomite.

Ostracon In epigraphy, material of little value, most often a potsherd on which there is some writing.

Overstriking An operation in which a coin is given a second legal design.

Paradeisos A Persian royal forest reserve.

Patristics A specialized discipline in the study of the writings of the fathers of the church, who were participants in and privileged witnesses to doctrinal debates in the early church.

Pentateuch In biblical exegesis, the group of first five books of the Bible, which is given the name *Torah* ('Law') in the Jewish tradition and is its foundational text.

Philisto-Arabian (Coins) This is the usual name for all coins of the Persian period discovered in the Near East (mainly in Palestine, Egypt and Arabia), closely or remotely resembling a Greek model; most of the time we know neither their date nor precise place of issue.

Positivism Starting from an idea of the experimental sciences already outdated about 1870, positivist history considers as scientific an inductive process based on an absolute empiricism: only documents subjected to a rigorous criticism permit the rediscovery of historical facts.

Proxeny In the Greek world, an honour that a city conferred by decree on a citizen of another city, which implied that this person, having become a proxenus, assumed in his city the defence of the interests of the city who had honoured him as well as of its citizens.

Relative Chronology Dating of any documents in reference to other documents that cannot be dated in an absolute chronology*; in archaeology, for example, the study of stratigraphy* makes it possible to obtain typological sequences of material (regional preferably) and to establish corpora arranged according to their chronological succession.

Remote Detection A technique using pictures taken from planes or satellites, utilized in archaeology as a geographical tool, to direct prospecting on the ground, to differentiate the environment of the uncovered sites and to propose components of an explanation on their distribution and typology.

Reorientation According to the typology of the syncretisms established by P. Lévêque, a particular orientation taken by a religion in contact with a foreign religion.

Seriality In history, the placing of a historical event in a temporal series of homogeneous and comparable units, making it possible to measure developments by way of time intervals over a fairly long period. In epigraphy, putting any new inscription in a series with homogeneous and comparable wholes, making it possible better to bring about its identification.

Shekel Weights and currency in use in the Middle East since earliest Antiquity. Phoenician coins with this name weighed about 14 grams.

Simulation An operation consisting of the construction of a model, that is, an approximate representation of a real system, in order to study its performance; simulation is practised, among other fields, in geography and, by extension, in archaeology.

Standard Proportion of precious metal which, according to regulations, should go into the alloy of a piece of coinage.

Stratigraphy In an archaeological excavation, the stratigraphic method consisting of clearing operations sufficiently controlled to distinguish the different layers of deposits and the various components in the occupation of the site, as a way to establish their sequence and, as a consequence, their relative chronology*.

Structural Anthropology Without ruling out history as an investigative method, C. Lévi-Strauss considers anthropology as the approach to the mental structures common to all humanity.

Syria-Palestine A geographical term inherited from ancient authors, used commonly to designate in a general way the Levantine Near East, by referring to the two large topographical and climatic entities of the region.

Talent A Greek unit of weight of 6000 drachmas, with the drachma being the main weight and monetary unit in ancient Greece (4.3 grams).

Theory of Central Places A theory of spatial analysis which proposes geometric simulations* to optimize the allocation and the hierarchy of clusters of population and of activities according to their functions in a complex society, occupying a homogeneous space.

Tophet In the Punic world, a sacred enclosure used as a place of burial for children, whose burnt remains were placed in urns surmounted by a stele.

Transection, Strip-Transection In physical geography, a representative sample of a given space (valley, mountain, region) according to a transverse route, generally rectilinear, used to study that space and bring to light its diversity.

Treasure Hoards Treasure made up of coins stored away for a more or less long period, consisting in general of coins of great value that had not circulated a great deal.

Yahwist (Name) A name terminated with the divine name Yahweh.

*Chronology of the Reigns of the Achaemenid Kings (*approximate dates)

Cyrus II the Great (Kūrush)	549–530
Cambyses II (Kambūjiya)	530–522
Smerdis (Bardiya)	522
Darius I (Darayavahuss)	522–486
Xerxes I (Khshayārshā)	486–465
Artaxerxes I 'Longimanus' (Artakhshassā)	465–424
Xerxes II	424–423
Darius II Nothus	423–404
Artaxerxes II Mnemon	404–359
Artaxerxes III Ochus	359–338
Arses (Ārshā)	338–336
Darius III Codomanus	336–330

BIBLIOGRAPHY

Abel, F.M.
  1967   *Géographie de la Palestine* (2 vols.; Paris: Gabalda [1933, 1938]).
Abercrombie, J.R.
  1984   'Computer assisted Alignment of the Greek and Hebrew Texts: Program-
              ming Background', *Textus* 11: 125-39.
Aharoni, Y.
  1962   *Excavations at Ramat Rahel, Seasons 1959 and 1960* (Rome: Centro die
              studi Semitici).
  1964   *Excavations at Ramat Rahel, Seasons 1961 and 1962* (Rome: Centro die
              studi Semitici).
  1966   *The Land of the Bible: A Historical Geography* (London: Burns and Oates,
              2nd edn).
Akurgal, E.
  1966   *Orient und Okzident* (Baden-Baden).
Albright, W.F.
  1938   *The Present State of Syro-Palestinian Archaeology and the Bible* (New
              Haven).
  1962   *Archaeology of Palestine* (Harmondsworth: Penguin Books, rev. edn
              [1949]).
Alt, A.
  1934   'Die Rolle Samarias bei der Entstehung des Judentums', in *Festschrift O.
              Procksh zum 60. Geburtstag* (ed. F. Baumgärtel *et al.*; Leipzig: Deichert
              and Hinrichs): 5-38.
Anderson, W.P.
  1975   'Analysis of Pottery Forms from Sounding Y according to a Type Series',
              in *Sarepta: A Preliminary Report on the Iron Age* (ed. J.B. Pritchard;
              Philadelphia: The University Museum): 53-70 and 106-14.
  1987   'The Kilns and Workshops of Sarepta (Sarafand, Lebanon): Remnants of a
              Phoenician Ceramic Industry', *Berytus* 35: 41-66.
Aron, R.
  1938   *Introduction à la philosophie de l'histoire: Essai sur les limites de
              l'objectivité historique* (Paris).
Auda, Y.
  1987   'A l'intersection des domaines de l'analyse des données et des sciences
              humaines: un tour d'horizon', *Archéologues et ordinateurs* 11: 3-15.
Avigad, N.
  1970   'Ammonite and Moabite Seals', in *Near Eastern Archaeology in the
              Twentieth Century: Essays in Honor of N. Glueck* (ed. J.A. Sanders; New
              York: Doubleday): 284-95.

1976   *Bullae and Seals from a Post-Exilic Judean Archive* (Qedem, 4; Jerusalem: Hebrew University).

1988   'Hebrew Seals and Sealings and their Significance for Biblical Research', in J.A. Emerton (ed.), *Congress Volume Jerusalem* (VTSup, 40; Leiden: Brill): 7-16.

Avi-Yonah, M.

1975–78   *Encyclopedia of the Holy Land* (4 vols.; London: Oxford University Press).

Babelon, E.

1910   *Traité des monnaies grecques et romaines*, II.2 (Paris).

Bajard, J.

1986   'Répertoire analytique des centres de traitement automatique de la Bible', in Müller (ed.) 1986: 81-85.

Bajard, J., and G. Servais

1986   'Présentation', in Müller (ed.) 1986: 9-11.

Balfet, H.

1980   'A propos du métier de l'argile: Exemple de dialogue entre ethnologie et archéologie', in Barrelet (ed.) 1980: 71-82.

Baly, D.

1957   *The Geography of the Bible: A Study in Historical Geography* (New York: Harper & Row).

Barraclough, G.

1980   *Tendances actuelles de l'histoire* (Paris).

Barrelet, M.T. (ed.)

1980   *L'archéologie de l'Iraq du début de l'époque néolithique à 333 avoat notre ère: Perspectives et limites de l'interprétation anthropologique des documents* (Paris: CNRS).

Barrois, A., and B. Carriere

1927   'Fouilles de l'Ecole archéologique française de Jérusalem effectuées à Neirab du 24 septembre au 5 novembre 1926', *Syria* 8: 126-42 and 201-12.

Baslez, M.F., and F. Briquel

1991   'De l'oral à l'écrit: le bilinguisme des phéniciens en Grèce', in C. Baurain *et al.* (eds.) 1991: 371-86.

Bastide, R.

1960   *Les religions africaines du Brésil* (Paris: Presses Universitaires de France).

1968   'Problèmes de l'entrecroisement des civilisations et de leurs oeuvres', in Gurvitch (ed.) 1968: II, 315-30.

Baurain, C., *et al.* (eds.)

1991   *Phoinikeia Grammata: Lire et écrire en méditerranée: IXe Colloque du Groupe du contact interuniversitaire d'études phéniciennes et puniques. Liège, 15-18 novembre 1989* (Namur: Société des études classiques).

Beeching, A.

1990   'Le programme d'archéologie spatiale: Culture et milieu des premiers paysans de la moyenne vallée du Rhône', in *Archéologie et Espaces: Xe Rencontre internationale d'Archéologie et d'Histoire, Antibes,* octobre 1989 (juan-les-Pins: APDCA): 137-55.

Bengtson, H. (ed.)
1968   *The Greeks and the Persians* (New York, 2nd edn [1965]).
Benichou-Safar, H.
1987   'Note sur le contenu des urnes du tophet de Carthage', in *Atti del IIo Congresso Internazionale di Studi Fenici e Punici, Roma, 9-14 nov. 1987.*
Ben-Tor, A., and Y. Portugali
1987   *Tell Qiri: A Village in the Jezreel Valley. Report of the Archaeological Excavations 1975–77* (Qedem, 24; Jerusalem in Hebrew University).
Binford, L.R., and S.R. Binford
1966   'A Preliminary Analysis of Functional Variability in the Mousterian of Levallois Facies', *AmA* 68: 238-95.
Bliss, F.J.
1898   *A Mound of Many Cities* (London: Palestine Exploration Fund).
Boardman, J.
1959   'Greek Potters at Al Mina?', *AnSt* 9: 163-69.
Bondi, S.F.
1978   'Note sull' economia fenicia. I. Impressa privata e ruolo dello stato', *Egitto e vicino Oriente* 1: 139-49.
Bordreuil, P.
1986   *Catalogue des sceaux ouest-sémitiques inscrits* (Paris: Bibliothèque Nationale).
Bordreuil, P., and D. Pardee
1989   *La trouvaille épigraphique d'Ugarit. I. Concordance* (Paris: P. Geuthner).
Bothma, T.J.
1989   'Computerized syntactic Data Bases in the Semitic Languages', *Journal for Semitics* 1: 23-38.
Braemer, F.
1989   'Occupation du sol dans la région de Jerash aux âges du Bronze Récent et du Fer', in *Studies in the History and Archaeology of Jordan* (Amman: Department of Antiquities): 191-98.
Braudel, F.
1969   *Ecrits sur l'histoire* (Paris: Flammarion).
1982   *La Méditerranée et le monde méditerranéen à l'époque de Philippe II* (Paris: A. Colin, 5th edn).
Briant, P.
1982   *Rois, tributs et paysans* (Paris).
1987   'Institutions perses et histoire comparatiste dans l'historiographie grecque', in Sancisi-Weerdenburg and Kuhrt (eds.) 1988: II, 1-10.
1988   'Ethno-classe domina-te et populations soumises dans l'Empire achéménide: le cas de l'Egypte', in Sancisi-Weerdenburg and Kuhrt (eds.) 1988: 137-74.
1996   *Histoire de l'empire perse de Cyrus à Alexandre* (Paris: Fayard).
Brochier, J.E.
1990   'Des techniques géo-archéologiques au service de l'étude des paysages et de leur exploitation', in J.-L. Fiches and S.E. van der Leeuw (eds.), *Archéologie et espaces, X^e rencontre internationale d'archéologie et d'Histoire Antibes, October 1989* (Juan les Pins: APOCA): 453-72.

Bry, M. de
1982   'La vie et l'oeuvre de Félix de Saulcy', in *Félix de Saulcy (1807–1880) et la Terre Sainte* (Notes et Documents des Musées de France, 5; Paris: Musées Nationaux): 17-69.
Buttrey, T.V.
1979   'The Athenian Currency Law of 375/4 B.C.', in *Greek Numismatics and Archaeology: Essays in Honor of M. Thompson* (Wetteren): 33-45.
Cagni, L.
1990   'Considérations sur les textes babyloniens de Neirab près d'Alep', *Trans* 2: 169-86.
Callataÿ, F. de
1984   'A propos du volume des émissions monétaires dans l'Antiquité', *RBNS* 130: 37-48.
Calmeyer, P.
1990   'Die sogennante fünfte Satrapie bei Herodot', *Trans* 3: 109-30.
Carena, O.
1989   *History of the Near Eastern Historiography and its Problems: 1852–1985. I. 1852–1945* (Neukirchen–Vluyn: Neukirchener Verlag).
Carter, G.F.
1983   'A Simplified Method for Calculating the Original Number of Dies from Die-Link Statistics', *ANSMN* 28: 195-206.
Certeau, M. de
1974   'L'opération historique', in Le Goff and Nora (eds.) 1974: I, 3-41.
Chaunu, P.
1978   *Histoire quantitative, histoire sérielle* (Paris: SEDES).
Chelhod, J., *et al.*
1984   *L'Arabie du Sud, histoire et civilisation. I. Le peuple yéménite et ses racines* (Paris: Maisonneuve et Larose).
Chenhall, R.G.
1981   'Computerized Data Bank Management', in *Data Bank Applications in Archaeology* (Tucson: S.W. Gaines): 1-8.
Claassen, W.T.
1987   'A Research Unit for Computer Applications to the Language and Text of the Old Testament', *JNSL* 1: 23-38.
1988   'Computer-assisted Methods and the Text and Language of the Old Testament: An Overview', in *idem* (ed.), *Text and Context: Old Testament and Semitic Studies for F.C. Fensham* (JSOTSup, 48; Sheffield: JSOT Press): 283-99.
Clerc, M.
1900   'De l'histoire considérée comme science', Discours de réception, in *Mémoires de l'Académie de Marseille* (Marseille).
Cleuziou, S., and J.P. Demoule
1980   'Situation de l'archéologie théorique', *Nouvelles de l'archéologie* 3: 7-15.
Collombier, A.M.
1987   'Modifications des lignes de rivage et ports antiques de Chypre: Etat de la question', in *Déplacements des lignes de rivage en Méditerranée* (Paris: CNRS): 159-72.

1988   'Harbour or Harbours of Kition on Southern Coastal Cyprus', in Raban
            (ed.) 1988: 35-46.

Condurachi, E.

1980   'Un exemple d'interculturalité: Le sud-est européen au premier millénaire
            avant notre ère', *Diogène (F)* 111: 116-41.

Crüsemann, F.

1989   'Le Pentateuque, une Tora. Prolégomènes et interprétations de sa forme
            finale', in de Pury (ed.), 1989: 339-60.

Dalongeville, R., and P. Sanlaville

1980   'Les changements de la ligne de rivage en Méditerranée orientale à l'époque
            historique: Exemple de la côte levantine', in *Salamine de Chypre* (Paris:
            Edition de Boccard): 19-32.

Dandamaev, M.A.

1976   *Persien unter den Achämeniden (6. Jahrhundert v. Chr.)* (Wiesbaden:
            L. Reichert).

Dar, S.H.

1986   *Landscape and Pattern: An Archaeological Survey of Samaria 800
            B.C.E.–636 C.E.* (British Archaeological Reports, International Series, 308;
            Oxford: Oxford University Press).

Delavault, B., and A. Lemaire

1979   'Les inscriptions phéniciennes de Palestine', *RSF* 7: 2-19.

Delcor, M.

1974   'De l'origine de quelques termes relatifs au vin en hébreu biblique et dans
            les langues voisines', in *Actes du premier Congrès international de linguis-
            tique sémitique et chamito-sémitique* (ed. A. Caquot and D. Cohen; Paris:
            Mouton): 223-33.

Dhorme, E.

1928   'Les tablettes babyloniennes de Neirab', *RA* 25: 53-82.

Djindjian, F.

1985   'La sériation en archéologie: un état de l'art. Méthodes et applications', in
            Panorama des traitements de données en archéologie (ed. H. Ducasse;
            Valbonne: CNRS/CRA): 9-46.

1991   *Méthodes pour l'archéologie* (Paris: Armand Colin).

Dothan, M.

1985   'Phoenician Inscription from 'Akko', *IEJ* 35: 81-94.

Dothan, T., and S. Gitin

1987   'The Rise and Fall of Ekron of the Philistines: Recent Excavations at an
            Urban Border Site', *BA* 50: 197-222.

1988   'Kh. El-Muqannah (Miqné-Ekron) 1985–1986', *RB* 95: 228-39.

Droysen, J.G.

1833   *Geschichte des Hellenismus* (Berlin).

Dunand, M.

1926   'Sondages archéologiques effectués à Bostan ech-Cheikh, près Saïda',
            Syria 7: 1-8.

1964   'Rapport préliminaire sur les fouilles de Byblos en 1962', *BMB* 17: 29-35.

1966   'Rapport préliminaire sur les fouilles de Byblos en 1964', *BMB* 19: 95-
            101.

Dupront, A.
1974   'La religion: Anthropologie religieuse', in Le Goff and Nora (eds.), 1974,
            II, 142-83.

Durand, J.L.
1984   'Le faire et le dire, vers une anthropologie des gestes iconiques', *History
            and Anthropology* 1: 29-42.
Dussaud, R.
1927   *Topographie historique de la Syrie antique et médiévale* (Paris: P. Geuth-
            ner).
1931   *La Syrie antique et médiévale illustrée* (Paris: P. Geuthner).
Elayi, J.
1978-79   'Le rôle de l'oracle de Delphes dans le conflit gréco-perse d'après les
            Histoires d'Hérodote I–II', *IrAnt* 13: 93-118; 14: 67-151.
1983   'Les monnaies de Byblos au sphinx et au faucon', *RSF* 11: 5-17.
1984   'Les symboles de la puissance militaire sur les monnaies de Byblos', *RN* 26
            (6th series): 40-47.
1986   'Un isolat de culture phénicienne', *RSF* 14: 113-15.
1987a   'Al-Mina sur l'Oronte à l'époque perse', in *Studia Phoenicia* (Leuven:
            Peeters): V, 249-66.
1987b   *Recherches sur les cités phéniciennes à l'époque perse* (AIONSup, 51;
            Naples).
1988   *Pénétration grecque en Phénicie sous l'empire perse* (Nancy: Presses
            Universitaires).
1989   *Sidon, eité autonome de l'empire perse* (Paris: Idéaphane).
1990a   'Point de vue sur les études phéniciennes', *BaghM* 21: 457-59.
1990b   'Tripoli (Liban) à l'époque perse', *Trans* 2: 59-72.
1990c   *L'économie de cités phéniciennes sous l'Empire perse* (AIONSup, 62:
            Naples).
1991a   'Remarques méthodologiques sur l'étude paléographique des légendes
            monétaires phéniciennes', in C. Baurain *et al.* (eds.) 1991: 187-200.
1991b   'Réflexion sur la place de l'histoire dans la recherche sur la Trans-
            euphratène achéménide', *Trans* 4: 73-80.
Elayi, J., and A.G. Elayi
1988   'Abbreviations and Numbers on Phoenician Prealexandrine Coinages: the
            Sidonian Example', *NAC* 17: 27-36.
1989   *La monnaie à travers les âges* (Paris: Ed. Idéaphane).
1993   *Trésors de monnaies phéniciennes et circulation monétaire (V^e-Ve s. av. J-
            C.)* (Supplements to Transeuphratène, 1; Paris: Gabalda).
Elayi, J., and Haykal M.R.,
1996   *Nouvelles découvertes sur les usages funéraires des Phéniciens d'Arwad*
            (Supplements to Transeuphratène, 4; Paris: Gabalda).
Elayi, J., and A. Lemaire
1989   'Numismatique', *Trans* 1: 155-64.
1990   'Les petites monnaies de Tyr au dauphin avec inscription', *NAC* 18: 99-
            116.
Fales, F.M.
1973   'Remarks on the Neirab Texts', *OrAnt* 12: 131-42.

Febvre, L.
1946   'Face au vent, manifeste des Annales nouvelles', *Annales* 1: 1-8.
1953   *Combats pour l'histoire* (Paris).
1954a   'Marc Bloch: dix ans après', *Annales* 9: 145-47.
1954b   'Moyen âge et Réforme, ou du pouvoir des étiquettes en histoire', *RHPhR* 34: 198-208.
1954c   'Souvenirs de Jules Bloch', *Annales* 9: 108-109.
Finkelstein, I.
1981   'Israelite and Hellenistic Farms in the Foothills of the Yarkon Basin', *ErIs* 15: 331-48.
Firmin, G.
1986   'Une banque de données sur les manuscrits de la Bible', in Müller (ed.) 1986: 402-403.
Forbes, A.D.
1987   'Syntactic Sequences in the Hebrew Bible', in E.W. Conrad and E.G. Newing (eds.), *Perspectives on Language and Text* (Winona Lake: Eisenbrauns): 59-70.
Franken, H.J.
1969   *Excavations at Tell Deir 'Allā I* (Leiden: E.J. Brill).
Frei, P., and K. Koch
1984   *Reichsidee und Reichsorganisation im Perserreich* (OBO, 55; Fribourg, Switzerland).
Frost, H.
1973   'The Offshore Island Harbour at Sidon and Other Phoenician Sites in the Light of New Dating Evidence', *IJNA* 2: 75-94.
Fugmann, E.
1958   *Hama: Fouilles et recherches de la Fondation Carlsberg, 1931–38. L'architecture des périodes pré-hellénistiques* (Copenhagen: Nationalmuseet).
Fulco, W.J., and F. Zayadine
1981   'Coins from Samaria-Sebaste', *ADAJ* 25: 197-226.
Furet, F.
1974   'Le quantitatif en histoire', in Le Goff and Nora (eds.) 1974: I, 42-61.
Galling, K.
1937   'Syrien in der Politik der Achämeniden bis 448 v. Chr.', *AO* 36: 3-4.
Gardin, J.C.
1955   'Problèmes de la documentation', *Diog(F)* 11: 107-24.
1958   'Four Codes for the Description of Artifacts: An Essay in Archaeological Techniques and Theory', *AmA* 60: 335-57.
Gardin, J.C., *et al.*
1987   *Systèmes experts et sciences humaines: Le cas de l'archéologie* (Paris: Egrolles).
Garelli, P.
1964   *L'Assyriologie* (Paris: Presses Universitaires de France).
Gates, M.H.
1988   'Dialogues between Ancient Near Eastern Texts and the Archaeological Records: Test Cases from Bronze Age Syria', *BASOR* 270: 63-91.
Gelb, I.
1980   *Computer-Aided Analysis of Amorite* (Chicago: Oriental Institute).

Gibert, P.
1979   *Hermann Gunkel (1862–1932) et les légendes de la Bible* (Paris: Flammarion).

Gordon, R.L.
1987   'Notes on some Sites in the Lower Wadi ez-Zerqa and Wadi Ragib', *ZDPV* 103: 67-77.

Gran-Aymerich, E., and J. Gran-Aymerich
1987a   'Charles Clermont-Ganneau', *Archéologia* 222: 71-79.
1987b   'Ernest Renan', *Archéologia* 224: 71-79.

Greenfield, J.C.
1985   'A Group of Phoenician City Seals', *IEJ* 35: 129-34.

Griffiths, A.
1987   'Democedes of Croton: A Greek Doctor at Darius' Court', in Sancisi-Weerdenburg and Kuhrt (eds.) 1987: 37-51.

Gubel, E.
1990   'Tell Kazel (Ṣumur/Simyra) à l'époque perse: Résultats préliminaires des trois premières campagnes de fouilles de l'Université Américaine de Beyrouth (1985-1987)', *Trans* 2: 37-50.

Gunkel, H.
1910   *Die Sagen der Genesis* (Göttingen).
1917   *Das Märchen im Alten Testament* (Tübingen).

Gurvitch, G. (ed.)
1968   *Traité de sociologie* (2 vols.; Paris: Presses Universitaires de France).

Guthe, H.
1911   *Bibelatlas* (Leipzig).

Hackens, T., and C.H. Carcassonne (eds.)
1983   *Pact 5: Statistique et Numismatique* (Louvain-la-Neuve: Séminaire de Numismatique Marcel Hoc).

Hackens, T., and G. Moucharte (eds.)
1992   *Numismatique et histoire économique phéniciennes et puniques* (Studia Phoenicia, 9; Numismatica Lovaniensa, 9; Louvain-la-Neuve: Seminaire de Numismatique Marcel Hoc).

Hadidi, A.
1987   'An Ammonite Tomb at Amman', *Levant* 19: 101-20.

Hamdy Bey, O., and T.H. Reinach
1892   *Une nécropole royale à Sidon* (Paris).

Hart, S.
1986   'Some preliminary Thoughts on Settlement in Southern Edom', *Levant* 18: 51-58.

Hauptmann, A., *et al.*
1986   'Chronique archéologique: Feinan 1984', *RB* 93: 236-38.

Hauptmann, A., and G. Weisgerber
1987   'Archaeometallurgical and Mining-archaeological Investigations in the Area of Feinan, Wadi 'Arabah (Jordan)', *ADAJ* 31: 419-37.

Henige, D.
1986   'Comparative Chronology and the Ancient Near East: A Case for Symbiosis', *BASOR* 261: 57-68.

Herion, G.A.
1986   'The Impact of Modern and Social Science Assumptions on the Recon-
          structions of Israelite History', *JSOT* 34: 3-33.
1987   'Sociological and Anthropological Methods in Old Testament Study', in
          Old Testament Essays 5: 43-64.
Herrenschmidt, C.L.
1976   'Désignation de l'empire et concepts politiques de Darius Ier d'après ses
          inscriptions en vieux-perse', *StIr* 5: 33-65.
1987   'Notes sur la parenté chez les perses au début de l'empire achéménide', in
          Sancisi-Weerdenburg and Kuhrt (eds.) 1987: 53-67.
Herskovits, M.J.
1952   *Les bases de l'anthropologie culturelle* (Paris).
Herzfeld, E.
1968   *The Persian Empire* (Wiesbaden: F. Steiner).
Herzog, Z., *et al.*
1980   'Excavations at Tel Michal 1978–1979', *Tel Aviv* 7: 11-51.
1981   'Notes and News', *IEJ* 31: 120-21.
Herzog, Z., *et al.* (eds.)
1989   *Excavations at Tel Michal, Israel* (Minneapolis: University of Minnesota
          Press).
Hill, G.F.
1910   *A Catalogue of the Greek Coins in the British Museum: Greek Coins of
          Phoenicia* (London).
Hoftijzer, J., and K Jongeling
1995   *Dictionary of the North-West Semitic Inscriptions* (2 vols.; Leiden: E.J.
          Brill).
Homan, M.J.
1988   'Computer-Assisted Biblical Research', *Concordia Journal* 14: 150-57.
Homes-Fredericq, D., and J.B. Hennessy
1989   *Archaeology of Jordan* (Supplement to Akkadica, 7-8; Leuven: Peeters): I–
          II.
Hours, F.
1980   'L'informatique et les ordinateurs dans l'archéologie du Proche-Orient: Le
          point de vue d'un utilisateur', *Paléorient* 6: 9-20.
Hurter, S., and E. Pászthory
1984   'Archaischer Silberfund aus dem Antilibanon', in *Festschrift für L.
          Mildenberg* (Wetteren): 111-25.
Jidejian, N.
1971   *Sidon through the Ages* (Beyrouth).
Joannes, F.
1990   'Pouvoirs locaux et organisation du territoire en Babylonie achéménide',
          Trans 3: 173-90.
Jobling, D.
1987   'Sociological and Literary Approaches to the Bible: How Shall the Twain
          Meet?', *JSOT* 38: 85-93.

Kenyon, K.
    1957   *Digging up Jericho* (London).
    1981   *Excavations at Jericho*. III. *The Architecture and Stratigraphy of the Tell* (London: British School of Archaeology in Jerusalem).
Khalil, L.A.
    1986   'A Bronze Caryatid Censer from Amman', *Levant* 18: 103-10.
Knauf, E.A.
    1990   'The Persian Administration in Arabia', *Trans* 2: 201-17.
Kraay, C.M., and P.R.S. Moorey
    1968   'Two Fifth-Century Hoards from the Near East', *RN* 10: 181-235.
Kraft, R.A., and E. Tov
    1981   'Computer-assisted Tools for Septuagint Studies', *BIOSCS* 14: 22-23.
Kramer, S.N.
    1957   *L'histoire commence à Sumer* (Paris: Arthand); originally published as *From the Tablets of Sumer* (Indian Hills, 1956).
Kuhrt, A., and H. Sancisi-Weerdenburg
    1987   'Introduction', in Sancisi-Weerdenburg and Kuhrt (eds.) 1987: ix-xiii.
Lagrange, M.S.
    1989   'Sciences humaines et systèmes experts', in *Sciences historiques, sciences du passé et nouvelles technologies d'information: Bilan et évolution* (Lille: Université de Lille III): 173-84.
Lang, B.
    1983   'Old Testament and Anthropology: A Preliminary Bibliography', *Biblische Notizen* 20: 37-46.
Langlois, C.H.V., and C.H. Seignobos
    1898   *Introduction aux études historiques* (Paris).
Lecocq, P.
    1990   'Observations sur le sens du mot *dahyu* dans les inscriptions achéménides', *Trans* 3: 131-40.
    1997   *Les inscriptions de la Perse achéménide* (Paris: Gallimard).
Lefebvre, G.
    1978   *Réflexions sur l'histoire* (Paris: Maspéro).
    1980   *La recherche historique en France depuis 1965* (Paris: Presses Universitaires de France).
Le Goff, J., and P. Nora
    1974   'Présentation', in *idem* (eds.) 1974: I, ix-xiii.
Le Goff, J., and P. Nora (eds.)
    1974   *Faire de l'histoire* (3 vols.; Paris: Gallimard).
Lehmann, G.
    1996   *Untersuchungen zur späten Eisenzeit in Syrian und Libanon: Stratigraphie und Keramikformen zwischen ca. 720 bis 300 v. Chr.* (Münster: Ugarit-Verlag).
Lemaire, A.
    1977   *Inscriptions hébraïques. I. Les ostraca* (Paris: Cerf).
    1988   'Recherches actuelles sur les sceaux nord-ouest sémitiques', *VT* 38: 220-30.

1990 'Populations et territoires de la Palestine à l'époque perse', *Trans* 3: 31-74.

1996 *Nouvelles inscriptions araméennes d'Idumée au Musée d'Israel* (Supplements to Transeuphratène, 3; Paris: Gabalda).

Lemaire, A., and J.-M. Durand

1984 *Les inscriptions araméennes de Sfiré et l'Assyrie de Shamshi-Ilu* (Paris: Droz).

Lemaire, A., and J. Elayi

1987 'Graffitis monétaires ouest-sémitiques', in Hackens and Moucharte (eds.) 1992: 59-67.

Lemaire, A., and H. Lozachmeur

1990 'La Cilicie à l'époque perse: Recherches sur les pouvoirs locaux et l'organisation du territoire', *Trans* 3: 143-56.

Leuze, O.

1935 *Die Satrapieneinteilung in Syrien und im Zweiströmlande von 520–332* (Halle).

Lévêque, P.

1973 'Essai de typologie des syncrétismes', in *idem* (ed.), *Syncrétismes dans les religions grecque et romaine* (Paris: Presses Universitaires de France): 179-87.

Lévine, L.

1973 'A propos de la fondation de la Tour de Straton', *RB* 80: 75-81.

Lévi-Strauss, C.L.

1958 *Anthropologie structurale* (Paris: Plon).

1973 *Anthropologie structurale* (Paris: Plon): II.

Lipiński, E.

1975 *Studies in Aramaic Inscriptions and Onomastics* (Leuven: Leuven University Press): I.

1990 'Géographie linguistique de la Transeuphratène à l'époque achéménide', *Trans* 3: 95-108.

MacAlister, R.A.S.

1911–12 *The Excavations of Gezer: 1902–1905 and 1907–1909* (3 vols.; London) .

McClelland, Th.L.

1979 'Chronology of the "Philistine" Burials at Tell el-Far'ah (South)', *Journal of Field Archaeology* 6: 57-73.

McEwan, G.J.P.

1982 *The Late Babylonian Texts in the Royal Ontario Museum* (Toronto: Royal Ontario Museum).

Maggiani, A.

1972 'Aska eleivana', *SEt* 40: 183-87.

Manfredi, V.

1986 *La Strada dei Diecimila* (Milano).

Maraval, P.

1988 'Saint Jérôme et le pélerinage aux lieux saints de Palestine', in *Jérôme entre l'Occident et l'Orient* (ed. Y.M. Duval; Paris: Etudes Augustiniennes): 345-53.

Mayes, A.D.H.

1988 'Sociology and the Study of the Old Testament: Some Recent Writings', *IBS* 10: 178-91.

Mazzoni, S.
 1990   'La période perse à Tell Mardikh et dans sa région dans le cadre de l'Age
              du Fer en Syrie', *Trans* 2: 187-200.
Mele, A.
 1979   *Il commercio greco arcaico: Prexis ed emporie* (Cahiers du Centre Jean
              Bérard, 4; Naples).
Meyers, E.M., *et al.*
 1976   *Ancient Synagogue Excavations at Khirbet Shema', Upper Galilee Israel:
              1970–1972* (AASOR, 42; Durham).
 1981   'Preliminary Report on the 1980 Excavations at en-Nabratein, Israel',
              BASOR 244: 1-26.
Mildenberg, L.
 1990   'Gaza Mint Authorities in Persian Times: Preliminary Studies of the local
              Coinage in the Fifth Persian Satrapy. Part 4', *Trans* 2: 137-46.
 1992   'The Philisto-Arabian Coins: A Preview', in Hackens and Moucharte (eds.)
              1992: 33-40.
Millard, A.R.
 1991   'The Uses of the early Alphabets: Phoenician', in Baurain *et al.* (eds.)
              1991: 101-14.
Miller, J.M.
 1983   'Site Identification: A Problem Area in contemporary Biblical Scholarship',
              ZDPV 99: 119-29.
Momigliano, A.
 1979   *Sagesses barbares: Les limites de l'hellénisation* (Paris: Maspéro [1976]).
 1987   *Pagine ebraiche* (Turin).
Montet, P.
 1926   *Byblos et l'Egypte* (Paris: Presses Universitaires de France).
Montmollin, G. de
 1976   *L'influence sociale* (Paris).
Morel, J.P.
 1983   'Les relations économiques dans l'Occident grec', in *Modes de contacts et
              processus de transformation dans les sociétés anciennes* (CEFR, 67; Pisa:
              Scuola Normale Superiore; Rome: Ecole Française de Rome): 549-80.
Moscati, S.
 1968   *The World of the Phoenicians* (London: Weinfeld & Nicolson).
 1982   *L'enigma dei Fenici* (Milan).
Müller, C. (ed.)
 1986   *Bible et informatique: Le texte* (Travaux de linguistique quantitative, 37;
              Paris: Champion; Geneva: Slatkine).
Musti, D.
 1981   *L'economia in Grecia* (Rome-Bari).
Naster, P.
 1987   'Trésors de monnaies de Byblos (IV^{e}s.) trouvés à Byblos', in Hackens and
              Moucharte (eds.) 1992: 41-49.
Negev, A. (ed.)
 1986   *The Archaeological Encyclopedia of the Holy Land* (Jerusalem:
              Weidenfeld & Nicolson).

Nicolet-Pierre, H.
1986   'L'oiseau d'Athènes, d'Egypte en Bactriane: Quelques remarques sur
           l'usage d'un type monétaire à l'époque classique', in L. Kahil *et al.* (eds.)
           Iconographie classique et identités régionales (BCH Supplement, 14;
           Paris: Ecole Française d'Athènes): 365-75.
Nora, P.
1974   'Le retour de l'événement', in Le Goff and Nora (eds.) 1974: I, 210-28.
North, R.
1979   *A History of Biblical Map Making* (Beihefte zum Tübinger Atlas des
           vorderen Orients, B-32; Wiesbaden: L. Reichert).
Noth, M.
1960   *Überlieferungsgeschichte des Pentateuch* (Darmstadt, 2nd edn [1948]).
1967   *Überlieferungsgeschichtliche Studien* (Darmstadt, 3rd edn [1943]).
Parpola, S.
1987   *The Correspondence of Sargon II*: I (Helsinki: Helsinki University Press).
Parrot, A.
1981   *Sumer* (Paris: Gallimard [1960]).
Perrot, G., and C.H. Chipiez
1885   *Histoire de l'art dans l'Antiquité. III. Phénicie-Chypre* (Paris).
Petrie, W.M.F.
1891   *Tell el-Hesy (Lachish)* (London: Palestine Exploration Fund).
1900-1901 *The Royal Tombs of the Earliest Dynasties* (2 vols.; London: The Egypt
           Exploration Fund).
Pettinato, G.
1975   'I rapporti politici di Tiro con l'Assiria alla luce del "trattato tra Asarhaddon
           e Baal"', *RSF* 3: 145-60.
Pézard, M.
1931   *Qadesh: Mission archéologique à Tell Nebi Mend, 1921–1922* (BAH;
           Paris: P. Geuthner).
Pocock, D.F.
1975   'North and South in the Book of Genesis', in J.H.M. Beattie and R.G.
           Lienhardt (eds.), *Studies in Social Anthropology* (Oxford): 273-84.
Polzin, R.
1977   *Biblical Structuralism* (Philadelphia: Fortress Press; Missoula: Scholars
           Press).
Poppa, R.
1978   *Kamid el-Loz 2, der eisenzeitliche Friedhof: Befunde und Funde* (Bonn:
           Rudolph Habelt Verlag).
Popper, K.
1934   *Logik der Forschung* (Vienna).
Postgate, N.
1980   'Palm-trees, Reeds and Rushes in Iraq Ancient and Modern', in Barrelet
           (ed.) 1980: 99-109.
Postma, F., *et al.*
1983   *Exodus: Materials in Automatic Text Processing*: I–II (Amsterdam: Vu
           Boekhandel; Turnhout: Brepols).
Préaux, C.L.
1965   'Réflexions sur l'entité hellénistique', *CEq* 40: 129-39.

Pritchard, J.B.
 1985   *Tell es-Sa'idiyeh: Excavations on the Tell, 1964–1966* (University
           Museum Monograph, 60; Philadelphia: The University Museum).
Pury, A. de (ed.)
 1989   *Le Pentateuque en question* (Geneva: Labor et Fides).
Pury, A. de, and T.H. Römer
 1989   'Le Pentateuque en question: position du problème et brève histoire de la
           recherche', in de Pury (ed.) 1989: 9-80.
Pury, A. de, *et al.* (eds.)
 1996   *Isräel construit son histoire* (Geneva: Labor et Fides).
Raban, A.
 1983   'Submerged Sites of the Coast of Israel', in N.C. Flemming and P. Masters
           (eds.), *Coastlines and Archaeology* (London): 215-32.
Raban, A. (ed.)
 1988   *Archaeology of Coastal Changes* (British Archaeological Reports Inter-
           national Series, 404; Oxford: Oxford University Press).
Raben, J., and S.K. Burton
 1981   'Information Systems and Services in the Arts and Humanities', *Annual
           Review of Information Science and Technology* 16: 247-66.
Radday, Y.T.
 1973   *The Unity of Isaiah in the Light of Statistical Linguistics* (Hildesheim:
           H.A. Gerstenbergh).
Radday, Y.T., *et al.*
 1982   'Genesis, Wellhausen and the Computer', *ZAW* 94: 467-81.
Rainey, A.F.
 1969   'The Satrapy "Beyond the River"', *AJBA* 1: 51-78.
Rak, Y., *et al.*
 1976   'Evidence of Violence on Human Bones in Israel, First and Third Centuries
           CE', *PEQ* 108: 55-58.
Rappaport, U.
 1981   'The First Judean Coinage', *JJS* 32: 2-17.
Redfield, R., *et al.*
 1936   'Memorandum for the Study of Acculturation', *AmA* 38: 149-52.
Reich, R.
 1984   'The Identification of the "Sealed Karu of Egypt"', *IEJ* 34: 32-38.
Reisner, G.A., *et al.*
 1924   *Harvard Excavations at Samaria 1908–1910* (2 vols.; Cambridge, MA:
           Harvard University Press).
Rémond, R.
 1982   'Une nouvelle histoire politique', in *Des repères pour l'homme* (Paris:
           Seine): 43-45.
Renan, E.
 1864   *Mission de Phénicie* (Paris).
Rendtorff, R.
 1986   *Introduction to the Old Testament* (Neukirchen: Neukirchener Verlag
           [1983]).

Renfrew, C.
1977   'Production and Exchange in Early State Societies, the Evidence of Pot-
            tery', in D.P.S. Peacock (ed.), *Pottery and Early Commerce: Charac-
            terization and Trade in Roman and Later Ceramics* (London: Academic
            Press): 1-20.
Renimel, S.
1979   'Reconnaître l'espace archéologique', *Dossiers de l'Archéologie* 39: 7-21.
Reynaud, Commandant
1914   'Alexandre le Grand colonisateur', *La Revue Hebdomadaire* (11 April):
            195-212.
Ridgway, D.
1973   'The First Western Greeks: Campanian Coasts and Southern Etruria', in
            Greeks, Celts and Romans (London): 5-36.
Ringel, J.
1975   *Césarée de Palestine: Etude historique et archéologique* (Paris: Ophrys).
Robert, J., and L. Robert
1977   'Toponymie antique dans l'Anatolie', in *La toponymie antique*
            (Strasbourg: Centre de Recherche sur le Proche-Orient et la Grèce antique):
            11-63.
Robert, P.H. de
1984   'Approches sociologiques de l'ancien Israël', *RHPhR* 64: 403-406.
Robertson Smith, W.
1889   *Lectures on the Religion of the Semites* (London).
Robinson, E., and E. Smith
1856a   *Biblical Researches in Palestine and in the Adjacent Regions: A Journal of
            Travels in the Year 1838* (2 vols.; Boston).
1856b   *Later Biblical Researches in Palestine and in the Adjacent Regions: A
            Journal of Travels in the Year 1852* (Boston).
Robinson, E.S.G.
1937   'Coins from the Excavations at Al-Mina (1936)', *NumC* 17: 186-90.
Rodd, C.S.
1981   'On Applying a Sociological Theory to Biblical Studies', *JSOT* 19: 95-
            106.
Röllig, W.
1983   'The Phoenician Language: Remarks on the Present State of Research', in
            Atti del I° Congresso Internazionale di Studi Fenici e Punici (Rome:
            Consiglio Nazionale delle Ricerche): 375-86.
Rogerson, J.W.
1974   *Myth in the Old Testament* (BZAW; Berlin: de Gruyter).
1978   *Anthropology and the Old Testament* (Oxford: Basil Blackwell).
1983   'The Use of Sociology in Old Testament Studies', in *Congress Volume
            Salamanca* (VTSup, 36; Leiden: E.J. Brill): 245-56.
Rosen, B.
1986–87   'Wine and Oil Allocations in the Samaria Ostraca', *Tel Aviv* 13-14: 39-45,
            71-84.

Rosenthal, F.
1983   'Die Krise der Orientalistik', in F. Steppart (ed.), *XXI. Deutscher Orien-*
            talistentag vom 24. bis 29. März 1980 in Berlin: Vorträge (ZDMGSup, 5;
            Wiesbaden: F. Steiner): 10-21.

Rowton, M.B.
1980   'Pastoralism and the Periphery', in Barrelet (ed.) 1980: 291-301.

Runciman, W.G.
1983   *A Treatise on Social Theory.* I. *The Methodology of Social Theory*
            (Cambridge).

Ryan, N.S.
1988   'A Bibliography of Computer Applications and Quantitative Methods in
            Archaeology', in *Computer and Quantitative Methods in Archaeology*
            (British Archaeological Reports International Series, 446; Oxford: Oxford
            University Press): 3-27.

Sabloff, J.A. (ed.)
1981   *Simulations in Archaeology* (Albuquerque: University of New Mexico
            Press).

Sancisi-Weerdenburg, H., and A. Kuhrt (eds.)
1987   *Achaemenid History.* II. *The Greek Sources* (Leiden: Nederlands Instituut
            voor het Nabije Oosten).
1988   *Achaemenid History.* III. *Method and Theory* (Leiden: Nederlands Instituut
            voor het Nabije Oosten).

Sanlaville, P.
1978   'Note sur la géomorphologie de la presqu'île d'Ibn Hani (Syrie)', *Syria* 55:
            303-305.

Sapin, J.
1989a   'Recension du livre de R. Poppa, 1978', *Trans* 1: 181-84.
1989b   'Un domaine de la couronne dans la Trouée de Homs (Syrie): Origines et
            transformations de Tiglat-Phalazar III à Auguste', *Trans* 1: 21-47.

Sasson, J.
1981   'On Choosing Models for Recreating Israelite Pre-Monarchic History',
            JSOT 21: 3-24.

Schaeffer, C.F.A.
1931   'Les fouilles de Minet-el-Beida et de Ras-Shamra: Deuxième campagne
            (Printemps 1930)', *Syria* 12: 1-14.
1939a   *Ugaritica* (Paris: P. Geuthner): I.
1939b   'Une trouvaille de monnaies archaïques grecques à Ras Shamra', in
            Mélanges Syriens offerts à R. Dussaud (BAH, 30; Paris: P. Geuthner):
            461-87.

Schnapp, A.
1974   'L'archéologie', in *Faire de l'histoire* (Paris), II: 3-24.

Schumacher, G.
1908   *Tell el-Mutesellim: Bericht über die 1903 bis 1905 veranstalteten Aus-*
            grabungen (Leipzig).

Sellin, E.
1904   *Tell Ta'annek: Bericht über eine Ausgrabung in Palästina* (DAWW, 50;
            Vienna).

Sellin, E., and C. Watzinger
    1913   *Jericho: Die Ergebnisse der Ausgrabungen* (WVDOG, 22; Leipzig).
Smith, G.A.
    1966   *The Historical Geography of the Holy Land* (London: Collins [1894]).
Stern, E.
    1978   *Excavations at Tel Mevorakh (1973-1976). I. From the Iron Age to the Roman Period* (Qedem, 9; Jerusalem: The Hebrew University).
    1979   'Excavations at Tel Mevorakh are Prelude to Tell Dor Dig', *BARev* 5.1: 34-39.
    1982   *The Material Culture of the Land of the Bible in the Persian Period 538–332 BC* (Warminster: Aris & Philipps; Jerusalem: Israel Exploration Society [1973]).
    1990   'The Dor Province in the Persian Period in the Light of the Recent Excavations at Tel Dor', *Trans* 2: 147-56.
    1995   *Excavations at Dor: Final Report* (Qedem Reports; Jerusalem: Hebrew University Press): IA, IB.
Stern, E. *et al.* (eds.)
    1994   *The New Encyclopaedia of Archaeological Excavations in the Holy Land* (4 vols.; Jerusalem: Carta and the Israel Exploration Society).
Stolper, M.
    1987   'Belshunu the Satrap', in F. Rochberg-Halton (eds.), *Language, Literature, and History: Philological and Historical Studies Presented to Erica Reiner* (New Haven: American Oriental Society): 389-402.
    1989   'The Governor of Babylon and Across-the River in 486 BC', *JNES* 48: 283-305.
Strange, J.F.
    1981   'Using the Microcomputer in the Field: The Case of the Meiron Excavation Project', *Newsletter of the ASOR* 4: 8-11.
    1984   'Recent Computer Applications in Ancient Near Eastern Archaeology', in H.O. Thompson (ed.), *The Answers Lie Below: Essays in Honor of L.E. Toombs* (Lanham, MD: University Press of America): 129-46.
Stroud, R.S.
    1974   'An Athenian Law on Silver Coinage', Hesp 43: 157-88.
Stucky, R.
    1983   *Ras Shamra, Leukos Limen* (BAH, 110; Paris: P. Geuthner).
Suder, R.W.
    1984   *Hebrew Inscriptions: A Classified Bibliography* (London and Toronto: Associated University Press).
Szacki, J.
    1979   *History of Sociological Thought* (London).
Talstra, E.
    1981   'The Use of *ken* in Biblical Hebrew: A Case Study in Automatic Text Processing', *OTSt* 21: 228-39.
Thompson, M., *et al.*
    1973   *An Inventory of Greek Coin Hoards* (New York).
Toombs, L.E., and N.E. Wagner
    1971   *Pottery Coding Handbook* (Waterloo, ON).

Beyond the River

Tunca, O.
1987 'Quelques réflexions sur les banques de données archéologiques', *Archéolog* 2: 65-70.
Ungnad, A.
1940–41 'Keilinschriftliche Beiträge zum Buch Esra und Esther', *ZAW*: 240-44.
Vallat, F.
1972 'Deux inscriptions néo-élamites de Darius Ier (DSf et DSz)', *StIr* 1: 3-14.
Van Effenterre, H.
1965 'Acculturation et histoire ancienne', in *XIIᵉ Congrès international des sciences historiques* (Vienna): 37-44.
Vanel, C.H.
1967 'Six "ostraca" phéniciens trouvés au temple d'Echmoun, près de Saïda', *BMB* 20: 45-95.
Vervenne, M.
1981 'A Bibliography of Bible and Computer', in *Centre: Informatique et Bible* (Maredsous): 52-86.
Wallis, L.
1912 *Sociological Study of the Bible* (Chicago).
Watkin, H.J.
1987 'The Cypriot Surrender to Persia', *JHS* 107: 154-63.
Weber, M.
1952 *Ancient Judaism* (Glencoe, IL: The Free Press).
Weil, G.E.
1964–65 'Méthodologie de la codification des textes sémitiques servant aux recherches de lingistique quantitative sur ordinateur', *BIIRHT* 13: 115-33.
1986 'Massorah, Massorètes et ordinateurs: Les sources textuelles et les recherches automatisées', in Müller (ed.) 1986: 351-61.
Weissbach, F.H.
1911 *Die Keilinschriften der Achämeniden* (VAB, 3; Leipzig).
Wellhausen, J.
1883 *Prolegomena zur Geschichte Israels* (Berlin).
1899 *Die Composition des Hexateuchs und der historischen Bücher des Alten Testaments* (Berlin).
Wheeler, M.
1954 *Archaeology from the Earth* (London: Penguin Books).
Whitaker, R.E.
1972 *A Concordance of the Ugaritic Literature* (AOAT, 19; Cambridge, MA: Harvard University Press).
Wiesehöfer, J.
1978 *Der Aufstand Gaumatas und die Anfänge Dareios* (Bonn: R. Habelt): I.
Will, E.D., and C.L. Orrieux
1986 *Ioudaïsmos-Hellènismos: Essai sur le judaïsme judéen à l'époque hellénistique* (Nancy: Presses Universitaires).
Williamson, H.G.M.
1987 *Ezra and Nehemiah* (OTG; Sheffield: JSOT Press).
1988 'The Governors of Judah under the Persians', *TynBul* 39: 59-82.

Wilson, R.R.
1984 *Sociological Approaches to the Old Testament* (Philadelphia: Fortress Press).

Wirth, E.
1971 *Syrien: Eine geographische Landeskunde* (Darmstadt: Wissenschaftliche Buchgesellschaft).

Woolley, L.
1938 'The Excavations at Al-Mina, Sueidia I', *JHS* 58: 1-30, 140-50.
1964 *Un royaume oublié* (Paris); originally published as *A Forgotten Kingdom* (London, 1953).

Yadin, Y.
1966 *Masada: La dernière citadelle d'Israël* (Paris: Hachette).

Yassine, K.
1983 'Tell el-Mazar, Field I: The Summit. Preliminary Report', *ADAJ* 27: 495-513.

Young, T.C.
1980 '480/479 BC: A Persian Perspective', *IrAnt* 15: 219-39.

Zertal, A.
1990 'The Pahwah of Samaria (Northern Israel) during the Persian Period: Types of Settlement, Economy, History and New Discoveries', *Trans* 3: 9-30.

Zysberg, A.
1986 'Impact de l'informatique sur la recherche historique', *Courrier du CNRS* 65: 12-13.

INDEXES

INDEX OF REFERENCES

OLD TESTAMENT

OTHER ANCIENT SOURCES

INDEX OF AUTHORS